Also by Roger Ellman:

THE ORIGIN AND ITS MEANING

*ON THE ORIGIN OF THE UNIVERSE AND ITS MECHANICS,
THE MECHANISM AND ORIGIN OF INTELLIGENCE,
AND THE IMPLICATIONS FOR THE INDIVIDUAL AND SOCIETY*

ON THE NATURE OF MATTER

*THE ORIGIN OF THE UNIVERSE CREATED MATTER
FUNDAMENTALLY WAVE IN NATURE, NOT PARTICULATE*

THE PHILOSOPHIC PRINCIPLES OF RATIONAL BEING

*ANALYSIS AND UNDERSTANDING OF
REALITY, TRUTH, GOODNESS, JUSTICE, VIRTUE, BEAUTY,
HAPPINESS, LOVE, HUMAN NATURE, SOCIETY, GOVERNMENT,
EDUCATION, DETERMINISM,
FREE WILL, AND DEATH*

GRAVITICS

THE PHYSICS OF THE BEHAVIOR AND CONTROL OF GRAVITATION

THE TROUBLE WITH THE HUBBLE LAW

*AN ALTERNATIVE TREATMENT OF REDSHIFTS IS EVIDENCED BY
FOUR INDEPENDENT OBSERVED COSMOLOGICAL EFFECTS*

RESOLUTION OF THE "SPOOKY" PROBLEMS OF QUANTUM MECHANICS

> THE WAVE NATURE OF PARTICLES AND LIGHT RESOLVES EINSTEIN'S "SPOOKY" ENTANGLEMENT PROBLEM AND THE "WEIRDNESS" OF QUANTUM MECHANICS

———————   ———————

HOW TO TRAVEL TO AND EXPLORE MARS OR ALPHA CENTAURI B

> USE OF FUEL IS IMPRACTICAL FOR SPACE EXPLORATION. THE ONLY FUEL-FREE ACCELERATION FOR SPACE EXPLORATION IS BY CONTROLLED GRAVITATION, AND THAT IS THE ONLY MEANS FOR A FLYING VEHICLE TO EXPLORE A DISTANT PLANET AND THE ONLY MEANS OF TRAVEL TO IT.

———————   ———————

PRIME OBJECTIVE TIME, PRIME OBJECTIVE SPACE

> EINSTEIN IN HIS THEORIES OF RELATIVITY INSISTED
> ON TWO UNIVERSAL PRINCIPLES:
>
> 1 – THERE IS NO PRIME FRAME OF REFERENCE,
> 2 - AND LIKEWISE THERE IS NO PRIMARY TIME.
>
> Neither of those denying principles is correct.
> It is here shown that there is
> a prime frame of reference, objective space.
> And, it is here shown that there is
> one grand overall flow of time, objective time.
>
> *Correcting those denials removes their adverse affects*
> *on Science and on Society*

ROGER ELLMAN

Cataloging Data

Ellman, Roger (1932-   )

Prime Objective Time, Prime Objective Space

That, contrary to the position of the theory of relativity, there is an objective prime time and there is an objective prime frame of spatial reference.

PRIME OBJECTIVE TIME, PRIME OBJECTIVE SPACE

Copyright (c) Roger Ellman 2020

All rights reserved. This book may not be reproduced nor transmitted in any form nor by any means, electronic, mechanical or other including but not limited to photocopying, recording or by any information storage or retrieval system without written permission from the author, except for the inclusion of brief quotations in a review.

Library of Congress Control Number:2019913464

Published by: The-Origin Foundation, Inc.,
                   1401 Fountaingrove Pkwy.
                   Santa Rosa, CA 95403, USA

                   707-537-0257

                   http://www.The-Origin.org

ISBN: 9781691079070

CONTENTS

Notes re Author..	iii
Preface	v

The Problem

1. The Claim of Einstein's Relativity ……………...…………………	1

Prime Objective Space

2. The Origin of Space and Matter …………………………………….	5
3. The Form and Behavior of Matter ………………………………….	15
4. Motion and Relativity ………………………………………………	27
5. Prime Objective Space ……………………...………………………	39
6. The Effect of Prime Objective Space on Science ……………….….	49

Prime Objective Time

7. Absolute Prime Time ……………………...……………………….	59
8. The Adverse Social Effect of Subjective Space and Time …………..	65

Appendices

Appendix A – Why No Immediate Mutual Annihilation ………………..	75
Appendix A-1 – The Neutron …………………………………..	85
Appendix B – The Limitation of the Original Envelopes ………………..	97
Appendix C – Derivation of Coulomb's Law ……………………………	101
Appendix D – The Universal Exponential Decay	115
Appendix E – The 'Big Bang' Outward Cosmic Expansion …………….	133

ABOUT THE AUTHOR

The-Origin Foundation, Inc. is a non-profit organization founded to foster independent scientific, mathematical, and philosophical research.

The author of the present work, Roger Ellman is the General Director of the foundation.

Roger Ellman has published over fifty professional papers on topics ranging from physics, cosmology, and astrophysics to artificial intelligence and mathematics.

He has presented some of his papers to conferences of / at:

> The American Physical Society [APS], .
> The American Society for the Advancement of Science,
> Cambridge University, United Kingdom
> The Library of Alexandria, Egypt
> The Russian Academy of Natural Sciences, St Petersburg
> The Hungarian Academy of Sciences, Budapest
> A Science Conference in Shang Hai, China

He is author of seven books in addition to the present "Prime Objective Time, Prime Objective Space".

His education includes graduate studies at Stanford University after graduating from West Point, the United States Military Academy.

PREFACE

In the 19th Century the development of Maxwell's equations and of the behavior of light led to a new problem. While Newton's Laws and other physical behavior readily transformed from one frame of reference to another by means of simple linear transformations, the new equations of electromagnetic behavior would not so transform.

It was the physicist Lorentz who resolved the problem with his "Lorentz Transforms", but that created another problem. It created a universe in which the mass of objects, their length, and the time involved in their motions were all relative, depending on the velocity of the object being considered.

And, Einstein in his theories of relativity codified these into two principles:

1 - That there is no prime frame of reference in the universe, rather all frames are relative, and

2 - That likewise there is no prime objective time in the universe, rather all time is relative.

Neither of those conclusions or principles is correct.

It is here shown that there is a prime [not preferred, not special, not different] frame of reference, objective space [which corresponds to the frame of the Big Bang].

And, it is here shown that there is one grand overall flow of time, prime objective time.

That reality of prime space and time, regardless of how we may view them from our particular frame and our particular motion, has had a significant adverse affect on science's understanding of the behavior of material matter.

And making those corrections removes a major adverse affect on human society that resulted partly from science's past denial of absolute space and absolute time.

Science has always had a major effect on the thinking of society.
And, Einstein's denial of prime objective space and time,
the 20th Century's attribution of uncertainty and indeterminism,
and Quantum Mechanics with its denial of cause and effect,
have had an adverse effect on both the progress of scientific
understanding
and the behavior of human society.

———

Establishing Prime Objective Space and Time
serves to correct those adverse effects

SECTION 1

THE PROBLEM

The Claim of Einstein's Relativity

In the 19th Century the development of Maxwell's equations and of the behavior of light led to a new problem. While Newton's Laws and other physical behavior readily transformed from one frame of reference to another by means of simple linear transformations, the new equations of electromagnetic behavior would not so transform.

It was the physicist Lorentz who resolved the problem with his "Lorentz Transforms", but that created another problem. It created a universe in which the mass of objects, their length, and the time involved in their motions were all relative, depending on the velocity of the object being considered.

And, Einstein in his theories of relativity codified these into two principles:

1 - There is no prime frame of reference in the universe; rather, all frames are relative.

and

2 - Likewise there is no prime objective time in the universe; rather, all time is relative.

A PRIME FRAME OF REFERENCE

With regard to space, that is frames of reference, Einstein's claim that there was no prime frame of reference was not based on experimental or observational fact; rather, it was his [firmly believed] opinion.

Einstein's principle concern and the reason for his contention against there being any prime frame of reference was his confidence that the laws of physics and the fundamental constants involved in those laws had to be the same everywhere regardless of the frame of reference and must be the same throughout the entire universe.

To Einstein the designation of any frame of reference as "prime" automatically meant that that frame was, or could be, different from the other frames of reference, that it was or could be, dominant in some sense, that it could violate the universality of physical laws.

That reasoning, upon which the speed of light being a universal constant and upon which that speed of light being the same in every frame of reference depended, led Einstein to his position absolutely opposed to any prime frame of reference.

That decision was quite unfortunate for two reasons.

- First, it was a decision as to the nature of an aspect of material reality for which there was completely no evidence. Neither passive observation nor experimental evidence soundly support nor supported such an important decision.

- Second, that decision automatically closed any possibility of investigating physical effects related to behavior relative to a prime frame of reference. As will be seen further below, that forced erroneous scientific decisions in explanation of observed behaviors.

Or, in other words, that decision restrained and distorted the scientific work thereafter. The decision and its support and acceptance in the scientific community were an example of failure of scientific objectivity, a failure to adhere to the "scientific method".

A UNIVERSAL PRIME OBJECTIVE TIME

A third reason as to why the decision against a prime frame of reference was unfortunate is that it made it easier, it opened the way, to assert, with similar lack of evidence, lack of justification, that there is no prime objective time in the universe, rather that all time is relative.

The reasoning employed here depends, again, on the speed of light, c. The problem is that two observers of an event, the observers well separated from each other and in different frames of reference cannot agree on an observed event which is distant from each observer but in different amounts of distance. They cannot agree because for each observer the event occurred at a time different from that experienced by the other observer because the light carrying the information about the event takes different times to travel the different distances of the observers from the event.

OVERALL AN ADVERSE EFFECT ON HUMAN SOCIETY

Circumstances can occur in which the observers cannot even agree on the order or sequence in time of related events such as cause and effect. That situation is completely unacceptable. Cause and effect are fundamental to physics and to the functioning of the universe. Objective, not relative, time is essential to the Reliability of cause and effect.

The combination of Einstein's denial of a prime frame of reference and the denial of prime objective time, both starting in the 20th Century, led to great damage to human society.

That is because science on the large scale, that is science dealing with the fundamentals of reality and the universe, has always had and still has a major effect on the non-scientific - social - general philosophic thinking of that science's society and its leaders as follows.

The beginning of the scientific method and the work of scientists such as Copernicus and Galileo resulted in the new period of "The Age of Reason" and "The Enlightenment" – rationality and empiricism replacing dogma and faith.

The new developments that Newton introduced led directly to the concept of the "clockwork universe" and the strong belief in laws, order and regularity.

And, Einstein's denial of objective space and objective time coupled with the 20th Century's attribution of actual uncertainty or indeterminism to all physical objects beyond the original Heisenberg conception, and the advent of Quantum Mechanics with its denial of cause and effect, resulted in our contemporary outlook of a probabilistic reality with no certainty, everything relative with no firm truths.

And, upon that we can lay some of the responsibility for the horrors and tragedies of the 20th Century because that has created the attitude that truth is different for each person and each case, that it is what each individual perceives it to be -- that there is no objective reality, only the subjective reality as perceived by each individual.

> The great damage that such thinking does is the license that it gives. It gives license to create, choose, decide upon one's own "reality" and then act accordingly. Such thinking ultimately gives us war, rapine, holocausts.

To address or understand the problem of
Prime Objective Space
it is necessary to begin at the beginning, that is to begin with the creation of space, the origin of the universe, the Big Bang.

Why is that so? The only "real" beginning, the only possible origin is a beginning of absolute nothing, before there was anything. That is the only beginning that requires no justification, no explanation, no accounting for its "existence".

The positing of any alternative is merely a postponing of the problem to an earlier, still ultimately necessary, absolute nothing.

However, that beginning of absolute nothing requires explanation of how a universe, or anything at all, could arise from "nothing".

That develops as follows.

SECTION 2

PRIME OBJECTIVE SPACE

The Origin of Space and Matter

INTRODUCTION

In order to correctly understand the nature of space it is necessary to understand the nature of matter. To do that it is necessary to consider all of the applicable sources of information and data. There are two such sources:

- The behavior of matter in its various encountered circumstances, and
- The origin of matter – how and from what it came to be.

Causality or mechanism is apparent from observation and experience which show that every thing and every event has a cause, and that those causes are themselves the results of precedent causes, and *ad infinitum*. Defining and comprehending the causality or mechanism operating to produce any contended or proposed scientific truth is essential to authenticating or validating that truth.

HOW THE UNIVERSE'S MATTER CAME TO BE

We are confronted with an apparently insuperable problem. Before the universe there was nothing, absolute nothing. That is the starting point because it naturally occurs; it is the only starting point that requires no cause, no explanation nor justification for its existence. But, that starting point has two impediments to the universe, or anything, coming into existence from it. First is the problem of change from nothing to something without, at least initially, an infinite rate of change, which is impossible. Second is the problem of change from nothing to something without violating conservation, which must be maintained.

The analysis would appear to end at that point, end with the declaration that obviously there cannot be a universe and there is no universe. Except, of course, that we and the universe we inhabit clearly exist at least enough for us to investigate it. Therefore, a solution to the insuperable problem exists. That solution is as follows.

1 - THE PROBLEM OF INFINITE RATE OF CHANGE

To avoid a material infinity the rate of change at the moment of the change must have been finite. Rather than an instantaneous jump from nothing to something, no matter how small or "negligible" that something might have been, there had to be a gradual transition at a finite rate of change. Further, the rate of change of that rate of change, the change's second derivative, at that moment had to have been finite, and so on *ad infinitum* for all of the further derivatives.

That requirement means that the form of the change had to have been either a natural exponential or some form of sinusoid. That develops as follows, in which the sought form of the change will be the function $U(t)$ [the "U" for universe, of course].

To illustrate the problem consider the function

(2-1) $U(t) = 0 \qquad t < 0$
$U(t) = t^2 \qquad t = 0$ and $t > 0$

as a theoretical candidate for $U(t)$ at the beginning of the universe, which function is graphically depicted at the right.

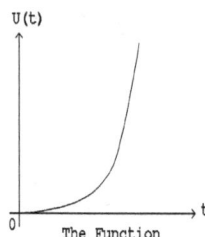

Its first derivative, also depicted graphically to the right, is

(2-2) $\dfrac{dU(t)}{dt} = 0 \qquad t < 0$

$\dfrac{dU(t)}{dt} = 2 \cdot t \qquad t > 0$

and is unstated for $t=0$ because $dU(t)/dt$ is not smooth there even though $U(t)$ "looks" smooth there.

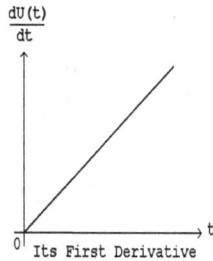

Figure 2-1a

Now, the second derivative depicted graphically to the right

(2-3) $\dfrac{d^2U(t)}{dt^2} = 0 \qquad t < 0$

$\dfrac{d^2U(t)}{dt^2} = 2 \qquad t > 0$

is clearly discontinuous at $t=0$, the instant of the beginning of the universe, where it instantaneously jumps from 0 to 2 as depicted to the right.

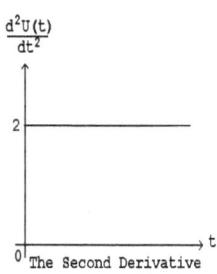

Figure 2-1b

The third derivative, which is the rate of change of the second derivative must be infinite at $t=0$ to produce the instantaneous jump from 0 to 2. Clearly, that cannot have happened in the real universe. It is such a condition which is unacceptable in a candidate function for $U(t)$ at the beginning of the universe.

The only way to avoid that condition of an infinite derivative somewhere along the line of successive further derivatives is to have a function with an endless family of finite, non-zero derivatives; that is, some derivatives may be zero at $t=0$ but there must always be further non-zero higher derivatives, which requires that the functional form of every derivative must be non-zero.

One can conceive theoretically of the idea of a function for which all derivatives are non-zero and no two are alike (in a general sense analogous to the pattern of digits in an irrational number), but it is not likely that such a function can exist. In any case the more certain and more simple way to achieve all non-zero derivatives is a repeating derivative function, the two simplest examples of which are as below.

2 – THE ORIGIN OF SPACE AND MATTER

(2-4) $$\frac{dU(t)}{dt} = \pm U(t) \quad \text{[First derivative = the original function]}$$

(2-5) $$\frac{d^2U(t)}{dt^2} = \pm U(t) \quad \text{[Second derivative = the original function]}$$

a. *Analysis of Repeating Derivative Functions*

Case (a): Functions Satisfying Equation 2-4

The function meeting this requirement is the natural exponential, ε^t.

(2-6) $$\varepsilon^t = 1 + t + \frac{t^2}{2!} + \frac{t^3}{3!} + \ldots$$

Taking the first derivative

(2-7) $$\frac{d[\varepsilon^t]}{dt} = 0 + 1 + \frac{2t}{2!} + \frac{3t^2}{3!} + \ldots$$
$$= 1 + t + \frac{t^2}{2!} + \frac{t^3}{3!} + \ldots = \varepsilon^t$$

so that the original function results as is required by equation *2-4*.

That is the prime case of a function that satisfies the requirement of all derivatives existing in functional form. In general those of this case are as equation *2-8*.

(2-8) $$U(t) = A \cdot \varepsilon^t$$

The function ε^t is not suitable for $U(t)$ at the beginning of the universe, however, because its value at $t=0$ is not zero. In fact it is zero only at $t = -\infty$. A function that might seem usable, however, would be

(2-9) $$U(t) = 0 \qquad t < 0 \text{ and } t = 0$$
$$U(t) = \varepsilon^t - 1 \qquad t > 0$$
$$= t + \frac{t^2}{2!} + \frac{t^3}{3!} + \ldots$$

which does have zero value at $t=0$ and otherwise meets the derivatives requirement sufficiently.

Cases (b) – (e): Functions Satisfying Equation (2-5)

Turning to functions that meet the requirement that the second derivative equal the original function per equation *2-5* there are four such functions.

(2-10)
Case (b): $$U(t) = 1 + \frac{t^2}{2!} + \frac{t^4}{4!} + \ldots$$

(2-11)
$$\text{Case (c):} \quad U(t) = 1 - \frac{t^2}{2!} + \frac{t^4}{4!} - \ldots$$

(2-12)
$$\text{Case (d):} \quad U(t) = t + \frac{t^3}{3!} + \frac{t^5}{5!} + \ldots$$

(2-13)
$$\text{Case (e):} \quad U(t) = t - \frac{t^3}{3!} + \frac{t^5}{5!} - \ldots$$

These five candidate functions can be described and summarized as their exponential equivalents as in Figure 2-2, below.

Case	Function	Name of Function	Candidate U(t)
(a)	ε^t	Natural exponential	$\varepsilon^t - 1$
(b)	$\dfrac{\varepsilon^t + \varepsilon^{-t}}{2}$	Hyperbolic cosine	$\text{Cosh}(t) - 1$
(c)	$\dfrac{\varepsilon^{i \cdot t} + \varepsilon^{-i \cdot t}}{2i}$	Cosine	$\text{Cos}(t) - 1$
(d)	$\dfrac{\varepsilon^t - \varepsilon^{-t}}{2}$	Hyperbolic sine	$\text{Sinh}(t)$
(e)	$\dfrac{\varepsilon^{i \cdot t} - \varepsilon^{-i \cdot t}}{2i}$	Sine	$\text{Sin}(t)$

Figure 2-2

The relationships in the table can be verified by substitution using the formula for ε^t as given in equation *2-6*, above. Cases *(b)* and *(c)* have the same problem that case *(a)* had, that the value of *U(t)* is not zero at *t=0*. Just as with case *(a)*, they would appear to become satisfactory if a constant, *1*, is subtracted from each of them.

These candidates all satisfactorily meet the requirement for a continuous family of derivatives so that the kind of unacceptable problem as encountered in the example of *U(t)=t²* at the beginning of this discussion is avoided. That is, all derivatives are finite. But, there are other requirements that the successful *U(t)* function must meet.

b. Using the Remaining Criteria to Select U(t)

Two other criteria must be met by the successful candidate function or functions:

- the function must not be open-ended, that is it cannot ever have an infinite amplitude, and

- the function must smoothly match the *U(t)=0* condition at *t=0*.

The first criterion eliminates cases *(a)*, *(b)* and *(d)* each of which goes to an infinite value of $U(t)$. To satisfy the second criterion the tangent to $U(t)$ at $t=0$ must be identical to the tangent to the function for $t < 0$, which is the horizontal *t-axis*. The condition is satisfied if the first derivative of $U(t)$ equals *zero* at $t=0$. Only cases *(b)* and *(c)* meet that requirement.

Therefore, the resulting form of $U(t)$, the only acceptable form, the only one that meets all of the requirements, is case *(c)*,

(2-14) $U(t) = [\cos(t) - 1]$ $t > 0$ and $t = 0$

   $U(t) = 0$ $t < 0$.

which is identical in form to the more usual and convenient equation *2-15*.

(2-15) $U(t) = U_0 \cdot [1 - \cos(2\pi \cdot f \cdot t)]$

in which an amplitude parameter, U_0, and a frequency parameter, *f*, have been added.

That the only possible form for the manner in which the universe began is a sinusoidal oscillatory form would seem to be very appropriate. Oscillations, waves, are ubiquitous in our universe from oceans, violin strings and pendulums to sound, light and electron orbits. That statement can also be validly inverted: Oscillations and waves are ubiquitous in our universe because the universe began from an initial such oscillatory form.

Every oscillation that we know in nature exhibits, and the very theory of oscillations in the abstract requires, that the oscillation consist of two aspects storing and exchanging the energy of the oscillation back and forth by means of a "flow". (With one aspect varying in oscillatory fashion then when that aspect decreases there must be some "place" for its energy to go, a place in which it is stored until it reappears in that aspect when it increases again. It cannot completely disappear or be lost because the oscillation would die. That "place" is the oscillation's second aspect and it obviously must vary in a manner related to the first aspect's variation, but with its energy storage in opposite phase.

A pendulum, for example, oscillates by the motion (flow) of its swinging mass between peak height in the gravitational field (potential energy) at each end of the swing and peak speed of motion (kinetic energy) at the mid-point between the ends of the swing. Then, what is the "flow" of the original oscillation at the start of the universe ? We do not know and likely will never know but we can give it a name, *Medium*, and we can investigate its characteristics and nature.

Such was the oscillation at the beginning of the universe except that at the first half cycle the energy was in only one form increasing from zero to its maximum. Then the second form began, similarly from zero to maximum, receiving and storing the energy of the first form as that gradually decreased in the second half cycle.

2 - THE PROBLEM OF CONSERVATION – "SOMETHING FROM NOTHING"

At this point, that is the universe having started from absolute nothing as an oscillation having the form of equation *2-15*, the maintaining of conservation, the avoiding of getting something from nothing, clearly could only happen in one manner:

There simultaneously had to have arisen an identical-in-form but opposite-in-amplitude oscillation so that the pair balanced out to the original net nothing, as in equation *2-16*.

(2-16) $U(t) = \pm U_0 \cdot [1 - \cos(2\pi \cdot f \cdot t)]$

There is no other way that violating the assured principle of conservation could have been avoided. The universe exists. It had to come into being from a prior nothing. That had to happen while avoiding an infinity of rate of change. Conservation had to be maintained. The universe began with the oscillation of equation *2-16*.

3. THE PROBLEM: WHY THAT OSCILLATION BEGAN AND WHAT IT WAS

a. Why That Beginning happened

A duration is the period of time that a particular state or set of conditions persists. The duration is terminated by a change, which change also initiates a new duration. In the universe change is ubiquitous. It is the constant and continuous stream of change that makes durations mensurable. Before the beginning of the universe a duration was in process even though it was not mensurable. The beginning of the universe was the first change ever and it terminated the original primal duration of absolute nothing.

The probability of the happening of such an event is extremely small. But the event was / is not impossible. Furthermore, in the absence of that event occurring there was an extremely large duration of opportunity in which that extremely small probability could operate. In the absence of the beginning the original duration would have been infinite and that infinite opportunity operated on by minute, but non-zero, probability results in absolute certainty. The beginning of the universe could not avoid eventually happening.

b. What That Beginning Oscillation Was

The starting point is the assumption that, when the primal nothing changed as a probabilistically inevitable interruption of what would otherwise have been an infinite duration of the primal nothing, the simplest or minimum conservation-maintaining interruption that could occur is what occurred. There are two reasons for this. Occam's Razor, calls for the simplest hypothesis as the most likely. More importantly, or perhaps the same thing, if an essentially spontaneous and extremely low probability event is to occur solely as an interruption of the duration of an otherwise absolute nothing, then very little interrupting event is needed; the barest minimum of something is sufficient to interrupt, to be a change in absolute nothing. There is no call, no reason for anything more. So, while the interruption could have been otherwise, it was probably as simple and minimum as possible.

Size or amount of time are of no meaning here because there is nothing to which they can be compared or by which they can be measured. Whatever amount of change occurred is what occurred. Whatever time it took, or went on for, whatever its oscillatory frequency was, is what happened. Twice as much or half as much have no meaning.

The following conclusions about the initial oscillatory $\pm U_0 \cdot [1 - \cos(2\pi \cdot f \cdot t)]$ form can now be reasonably obtained:

- clearly the universe of today must be an on-going evolved consequence of its beginning, of the initial oscillatory form;

2 – THE ORIGIN OF SPACE AND MATTER

- the frequency, f, of the sinusoidal oscillation was, and is, very large; and

- the nature of the change is one of concentration or density of the something that is oscillating.

- the oscillation was spherical, radially outward in all directions from its origin, because there was nothing to constrain it otherwise.

The frequency would have to be either very large or very small -- high enough so that it is not detected or noticed by us in every day life or so low that it appears to us as no change at all in our experience.

It has already been noted that the fact that the only possible form for the manner in which the universe began is a sinusoidal oscillatory form is very appropriate because oscillations, waves, are ubiquitous in our universe from oceans, violin strings and pendulums to sound, light and electron orbits. And it has been noted that that statement can be validly inverted: oscillations and waves are ubiquitous in our universe because the universe began from an initial such oscillatory form.

If the frequency of the initial oscillation were so small that it appears to us as no change at all it would completely eliminate oscillations playing any significant part in the behavior of the universe as we know it. Therefore, the frequency must have been very large, so rapid compared to our perception that we do not notice the oscillation at all.

The change can hardly be one of gross size if it is going on right now at high frequency as has just been concluded. One can conceive of the fundamental "substance", the "something" of the universe flashing into and out of existence from a zero to a maximum density or concentration in an oscillatory fashion at a rate so high that we neither detect nor notice it at all. But, it is not possible to entertain a concept of reality flashing from zero to full size, a size that includes ourselves and our environment, in such a fashion.

Actually, the reality that we know is not "flashing into and out of existence ...." Our reality is more the oscillation itself than what is oscillating and the continuing oscillation is our steady, constant reality.

Thus the interruption that gave us our universe was the starting of an *oscillation* that was *spherical*, present to us at a very high frequency and of $\pm U_0 \cdot [1 - Cos(2\pi \cdot f \cdot t)]$ form, of the density, as the variation will be hereafter referred to, of the *Medium*, as what it is that is oscillating will be hereafter referred to.

All of the discussion so far must apply to the "negative" oscillation, $-U(t)$, exactly as to the "positive" oscillation $+U(t)$ because the exact same reasoning as for $+U(t)$ applies to $-U(t)$ and, after all, they are not distinguishable in the discussion. The terms "+" and "-" are merely terms of convenience for two equal form opposite magnitude unknown things. We probably tend to think of our universe as the "+", but that is meaningless and irrelevant. There can be no objective designation of $+U(t)$ and $-U(t)$, no way to identify one versus the other. Both had to appear and our universe cannot avoid being the evolved result of both.

The universe that we know and exist in is the combined integrated result of both $+U(t)$ and $-U(t)$. The "+" and "-" electric charges of our universe [in both matter as for example in protons and electrons and in anti-matter as for example in negaprotons and positrons] must derive from that aspect of the beginning. (It is interesting to observe, also, that our universe being the integrated result of an initial beginning and its opposite relates to (presumably is the underlying cause of) the dialectical nature of reality, the ying and yang of oriental philosophy.)

The question of what the *Medium* is can only be answered in terms of its characteristics, what it does and how. Its characteristics are:

- a continuous entity, not a mass of "particles" nor anything having parts,

- simple and uniform throughout,

- of minimum tangibility or substantiality, not unlike the actuality of what we designate as "field" [electric, gravitational, etc].

4. THE PROBLEM: WHY DID THE EFFECTS OF EQUATION 2-16 NOT PROMPTLY CANCEL AND ON-GOING ABSOLUTE NOTHING RESUME ?

This is resolved in detail in Appendix A, "Why No Immediate Mutual Annihilation". Briefly, the initial structure was so unstable that it promptly exploded in that which we refer to as the "Big Bang" before annihilation could occur.

5. THE PROBLEM: IT HAS BEEN THOUGHT THAT THE UNIVERSE HAD TO START AT A POINT, A SINGULARITY. HOW COULD A DIMENSIONLESS POINT DELIVER A WHOLE UNIVERSE?

The sole reason for positing a point origin was to avoid an initial infinite rate of change. The gradualness of the `[1 - Cosine]` form resolves the problem of avoiding an infinite rate of change so that a point origin is no longer required.

The Big Bang "event horizon" problem and its relation to the development of variety in the universe has led to the hypothesis that there was an initial brief period of extremely rapid expansion called "inflation". That hypothesis has no supporting cause nor mechanism except its role in meeting the "event horizon" problem.

But with the need for a point origin eliminated the origin can have started per equation `2-16` at any size. There was no un-accounted-for period of "inflation". From estimates calculated of the number of particles in today's universe it has been determined that the initial, at the very first instant, the already "inflated"-size universe began. It was a highly concentrated volume of all of the mass and energy of the universe of about `40,000 km` radius.

That size is in terms of today's sizes [km]. For that event specific size is meaningless because there was nothing else to compare it to.

> *Next: How those Original Oscillations became the universe.*
> *Section 3 – The form and behavior of matter*

SECTION 3

The Form and Behavior of Matter

Section 2, *The Origin of Space and Matter* resolved the origin of the matter of the universe as follows.

The universe exists. It had to come into being from a prior nothing. That had to happen while avoiding an infinity of rate of change. Conservation had to be maintained. *Ergo* equation 2-16.

(2-16) $U(t) = \pm U_0 \cdot [1 - \cos(2\pi \cdot f \cdot t)]$

Thus the hypothesis is that the interruption that started our universe, the interruption of what would otherwise have been an infinite duration of the primordial absolute nothing, an interruption because an essentially infinite amount of opportunity operated on a non-zero though minute probability, was the starting of a matched pair of spherical oscillations:

- Present to us at a very high frequency,
- Of the general *[1 - Cosine]* form, and
- Together equal to the original nothing because of having matching amplitudes $+U_0$ and $-U_0$.

That analysis yielded an initial event, the origin oscillations, as in Figure 3-1. [All of the unavoidably planar depictions of the spherical oscillations are of the spherical phenomenon, interpretable as a radial versus time depiction.]

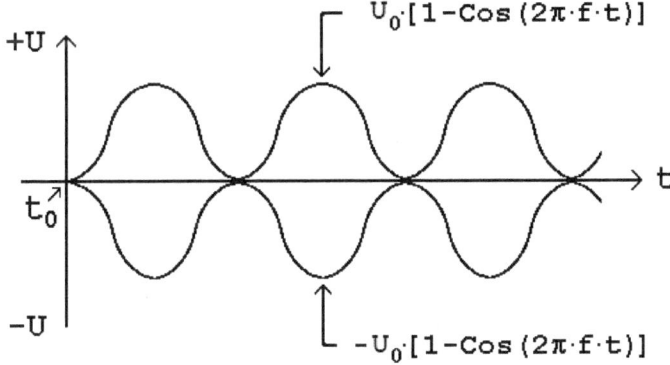

Figure 3-1

How the Original Oscillations Became the Universe

Examination of the waveform of Figure 3-1 reveals two problems. One, that it is an immediate mutual annihilation, will be dealt with shortly below. Of concern now is the second that an infinite rate of change still remains; the envelope of the oscillation has an infinite rate of change at $t=t_0$ as can be seen in Figure 3-2, below, which displays the envelope.

Viewed in a mathematical or graphical sense without any consideration of the physical reality represented, the envelope discontinuity at $t=t_0$ is not a difficulty. The only quantity that actually exists and is varying is the overall $U(t)$. The envelope is merely our perception of a characteristic of the waveform. The actual varying quantity, per Figure 3-1, has no discontinuity at $t=t_0$

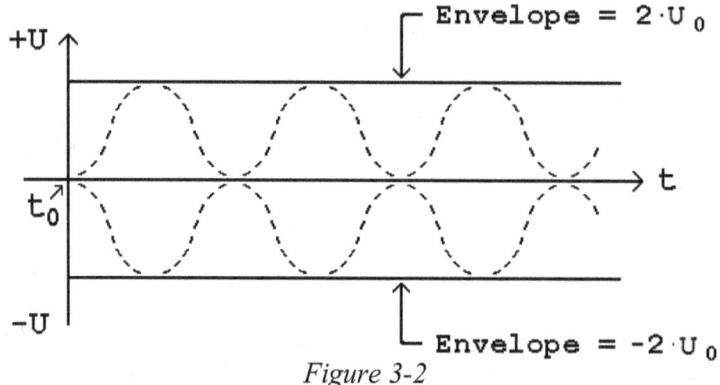

Figure 3-2

However, looking at the situation in a physical sense rather than purely mathematically, such oscillations as depicted in Figure 3-1 are all that there is to account for the effects which we call *energy*, *mass* and *charge*. Therefore, this *energy / mass / charge / oscillation* is something other than nothing. It is a physical reality that did not exist prior to the Origin. It can no more leap from zero to a finite non-zero amount than could the original $U(t)$ so leap.

That infinite rate of change in the amount of *energy / mass /charge* at $t=t_0$ is no more acceptable than was the infinite rate of change encountered in the original analysis of the beginning of the universe and it must be corrected by the same kind of reasoning as was then pursued: the envelope, also, had to originate as a *[1 - Cosine]* form of oscillation, which is the only form that avoids an infinite rate of change and matches the requirements of the situation.

That original envelope oscillation was at a lesser frequency than the original wave by the definition of a waveform envelope. If it were at a greater frequency then the roles (envelope and wave) would be reversed. If it were at the same frequency it would not act as an envelope and the infinity problem would remain. If we designate the envelope frequency as f_{env} and the frequency of the wave oscillation within the envelope as f_{wve} then the envelope would be of the following form.

(3-1) $\quad U_{env} = [1 - \cos(2\pi \cdot f_{env} \cdot t)]$

The wave is, as before, of the form

(3-2) $\quad U_{wve} = \pm U_0 \cdot [1 - \cos(2\pi \cdot f_{wve} \cdot t)]$

and the envelope modulating the wave is then

(3-3) $\quad U(t) = [U_{env}] \cdot [U_{wve}]$

$\qquad\qquad = \pm U_0 \cdot [1 - \cos(2\pi \cdot f_{env} \cdot t)] \cdot [1 - \cos(2\pi \cdot f_{wve} \cdot t)].$

That waveform appears in Figure 3-3.

However, the form of U(t) of equation 3-3 and Figure 3-3 still does not resolve the problem of an infinite rate of change at t_0. The *[1 - Cosine]* envelope is itself an oscillation that begins at t_0 with a sudden step from zero to its full amplitude. Figure 3-3 shows the first *2* cycles of the enveloped oscillation, which if only the envelope is considered, is a simple oscillation at the envelope frequency, even though visually, in the figure, it is only the trace of the peaks of the overall complex oscillation.

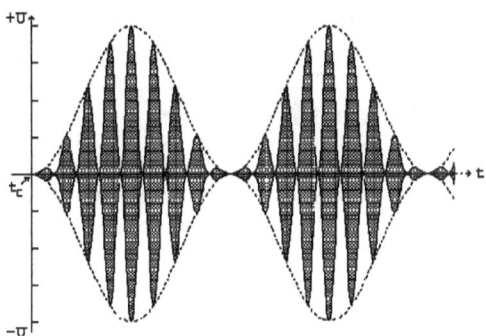

Figure 3-3

It is nevertheless *energy / mass / charge* that begins suddenly in its full amount at t_0 just as, in Figure 3-1, the oscillation of equation 3-1 begins at t_0. Therefore, it is again necessary to introduce an envelope of *[1 - Cosine]* form to prevent the infinite rate of change at t_0 in the prior envelope. That correction will in turn require still another such correction and so *ad infinitum*. An (apparently at this point) infinite string of envelopes thus results as a necessity of the situation.

The resulting U(t) then is

$$(3\text{-}4) \quad U(t) = \pm U_0 \cdot \prod_{i=1}^{i=\infty} \left[[1 - \cos(2\pi \cdot f_{env_i} \cdot t)] \right] \cdot \ldots$$

$$\ldots \cdot \left[[1 - \cos(2\pi \cdot f_{wve} \cdot t)] \right]$$

```
where the  ||  symbol (a large π, Greek "p")
means the product of the indicated factors.
```

While an envelope frequency must be less than the frequency of the wave that it modulates so that the various f_{env} must be less than f_{wve}, each successive envelope may be at the same frequency, as the prior. The reason is as follows.

If each envelope frequency must be different then each must be at least slightly smaller than the prior. With an infinite set of envelopes and only the frequency range from slightly less than that of the wave down to slightly above zero being available each successive envelope could only be at an infinitesimally lower frequency than its predecessor in any case. Infinitesimally less is essentially the same as identical.

Then how did other than an infinite string of envelopes come about ?

Each additional envelope factor in equation 3-4 results in a higher frequency content in the overall expression. That is, as each envelope is added the expansion of the exponentiated cosines expression into a sum of individual frequency cosine terms becomes longer and acquires higher frequency terms. But, the oscillation could not have had an actual component at infinite frequency. The real universe original $U(t)$ had an enormous set of envelopes but not an infinite set; they were "cut off" at some point.

The *Medium* of these oscillations being the only reality and, therefore, being what sets the limit on the speed of light with which we are familiar, the *Medium* also sets a limit on the highest frequency / lowest wavelength waves that can propagate. As a result the series of envelopes, of factors in equation 3-4, was limited to some finite but quite large amount. (See Appendix B, "The Limitation of the Original Envelopes").

This yields a revised $U(t)$, the original oscillation, the Cosmic Egg, as equation 3-5, below. N_0 is the number of envelopes, all at the same frequency, f_{env}.

$$(3-5) \qquad U(t) = \pm U_0 \cdot \left[1 - \cos[2\cdot\pi\cdot f_{env}\cdot t]\right]^{N_0} \cdot \left[1 - \cos[2\cdot\pi\cdot f_{wve}\cdot t]\right]$$

The waveform $[1 - \cos(x)]^n$ converges to an increasingly narrower peak as n increases, Figure 3-4, below. For very large *n*, that is very large N_0 of equation 3-5, the converging of the waveform into a single narrow peak proceeds to a momentary "spike" per cycle. Figure 3-5, below, shows the appearance of the waveform for extremely large *n*, that is for $n = N_0$ - what the waveform of the original "Cosmic Egg", the start of our universe, "looked like". (N_0 is found further below to be about 10^{84}.)

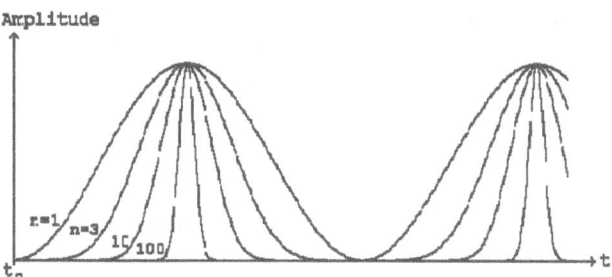

Figure 3-4 $[1 - \cos(x)]^n$ For n = 1, 3, 10, 100

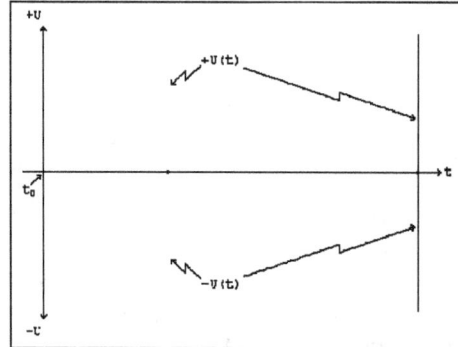

Figure 3-5
The U(t) "Cosmic Egg" WaveForm

This discussion of $U(t)$, the original oscillation the start of which was the start of the universe, has dealt so far only with the problems of the Origin, the problems of the transition from nothing to something. The something was, of course, the first instant of the entire universe. As such it must have contained in itself all of the *mass / energy / positive and negative charge* of the universe.

Figures 3-1, 3-3, and 3-5 all indicate that the original pair of oscillations, $+U$ and $-U$, should have immediately mutually annihilated, canceled out, reverted to the primal nothing. But, clearly that did not happen. The only explanation of that not happening is that each was unstable, so unstable that they exploded more immediately than they were able to mutually annihilate. They immediately proceeded to an immense explosion of energy and pieces of their oscillation, the event now called the "Big Bang". See Appendix A, "Why No Immediate Mutual Annihilation".

In terms of the $U(t)$ as depicted in Figure 3-5, the so immediate explosive decay undoubtedly occurred after only a minute portion, an infinitesimal portion, of the very first cycle had passed. It had to have been long before the first "spike". In that sense the initial event was very small, tenuous, hardly more than nothing because the instantaneous amplitude of $U(t)$ at that moment (the height of the curve above zero at that moment long before the first "spike") was also infinitesimal. It was hardly more than, essentially zero.

In that sense, the way that the universe started at all becomes a little more comprehensible. To avoid an infinite rate of change there was essentially almost no difference between "nothing", on-going absolute nothing, and the first infinitesimal moment of the original $U(t)$, the original oscillation.

Yet, it contained the entire universe.

THE FORM OF MATTER AS GENERATED BY THE "BIG BANG"

What did the "Cosmic Egg" explode into ? It could only explode into pieces of what it was made of, pieces of *[1 – cosine]* form spherical oscillations, pieces like equation 3-16, above.

Each oscillation is three-dimensional, thus spherical, because three dimensions is the minimum number that can involve space part of which is not its own boundary.

But, what did the "Cosmic Egg" explode into ? It primarily exploded into what we know our universe to mainly consist of: myriad protons - Hydrogen atom nuclei, and myriad electrons - maintaining overall charge neutrality with the protons, and the antimatter forms of both, negaprotons and positrons – maintaining conservation.

[Those might also be expected to have mutually annihilated but did not. Their survival rather than annihilation is analyzed in full in Appendix A, "Why No Immediate Mutual Annihilation". Suffice it here to observe that each product piece was initially ejected radially outward at extreme velocity and energy, on paths slightly diverging, such that initially annihilations could not occur.]

Then, what was the nature, the form of those product pieces that the "Cosmic Egg" exploded into ? Because of the two frequencies of $U(t)$, f_{wve} and f_{env}, and that the explosion source was of two equal but opposite polarities, $+U_0$ and $-U_0$, the "Big Bang" resulted in myriad pieces of four different forms of *[1 – cosine]* form spherical oscillations , Equations 3-6 .

$$(3-6) \quad U_{Form\ 1}(t) = +U_c \cdot [1 - \cos(2\pi \cdot f_{wve} \cdot t)] \quad \text{the proton}$$

$$U_{Form\ 2}(t) = -U_c \cdot [1 - \cos(2\pi \cdot f_{env} \cdot t)] \quad \text{the electron}$$

$$U_{Form\ 3}(t) = -U_c \cdot [1 - \cos(2\pi \cdot f_{wve} \cdot t)] \quad \text{the anti-proton}$$

$$U_{Form\ 4}(t) = +U_c \cdot [1 - \cos(2\pi \cdot f_{env} \cdot t)] \quad \text{the anti-electron}$$

Each of those has a specific value of its mass. Per the data provided by NIST, the National Institute of Standards and Technology those masses are:

(3-6a) ■ the proton and the antiproton $m_p = 1.672\ 621\ 898 \cdot 10^{-27}$ kg

■ the electron and the anti-electron $m_e = 9.109\ 383\ 56 \cdot 10^{-31}$ kg.

Using the mass-energy relationship, $m \cdot c^2 = h \cdot f$ the frequency, f, of those particles can be calculated. Those frequencies are:

(3-6b) ■ the proton and anti-proton: $f_{wve} = 2.268,731,818 \cdot 10^{23}$ hz

■ the electron and anti-electron: $f_{env} = 1.235,589,965 \cdot 10^{20}$ hz.

Finally, the mass of those four fundamental particles having now been resolved, their electric charge remains. They all have the same magnitude of their oscillation, $|U_c|$, which by default is the magnitude of their electric charge. [U_c is the particle oscillation amplitude per equation 3-6. U_0 is the original pre-explosion Big Bang oscillation amplitude.] The magnitude of the oscillation is in two opposite polarities; therefore clearly, where q is the fundamental electric charge per NIST, then:

(3-7) $q = 1.602,176,621 \times 10^{-19}$ coulombs

$+U_c$ corresponds to $+q$ [see * below]

$-U_c$ corresponds to $-q$ [see * below]

[* See in " C, Derivation of Coulomb's Law", at after equation C-22, the analysis:

"Understanding The Units of Charge and of Coulomb's Law"]

Judging by its result, the "Cosmic Egg" was not unlike an immense atom, a very unstable immense atom [as are all of the atomic species of atomic number exceeding 83 which the cosmic egg would have immensely exceeded]. Its "Big Bang" was <u>a kind of explosive nuclear radioactive decay</u> ultimately ending in the myriad stable elements of today's Periodic Table Of the Elements plus those with half lives long enough to be in detectable quantities today. Such decays follow a chain:

- From a heavy and complex composition,

- To various multiple less heavy less complex product pieces,

. . . . .

- Until they arrive at many multiple stable forms.

The vast majority of those resulting stable forms are the protons and electrons of the material world and their anti-particles. They are of the equation *3-6* form spherical oscillation, and will be referred to as *Spherical-Centers-of-Oscillation* or as *particles*

The rates of the decays are exponential, the decay [varying from some extremely rapid to some extremely slow] is described in terms of a "half life", the time it takes for half of the original material's decays to take place. Some of those "multiple less heavy

3 – THE FORM AND BEHAVIOR OF MATTER

less complex product pieces" having long half lives are present to us still today still decaying as what we term "radioactive" species.

THE FLOW FROM THE SPHERICAL-CENTERS-OF-OSCILLATION

The Particle "Core"

Consider a small individual particle such as a proton. Newton's law of gravitation expressed in terms of m_{source} and $m_{acted-on}$ and with both sides of the equation divided by $m_{acted-on}$ is, of course,

$$(3\text{-}7) \qquad a_{grav} = G \cdot \left[\frac{m_{source}}{d^2}\right]$$

However, mass and energy are equivalent, so that [using c = light speed and h = Planck's constant] a mass, m, is proportional to a frequency, f, that is characteristic of that mass. That is

$$(3\text{-}8) \qquad m \cdot c^2 = h \cdot f \quad \text{or} \quad f = [c^2/h] \cdot m$$

so that the m_{source} of equation 3-7 has a corresponding equivalent frequency, f_{source}.

That being the case, the gravitational acceleration, a_{grav}, can be expressed in terms of that frequency as the change, Δv, in the velocity, v, of the attracted mass per time period, T_{source}, of the oscillation at the corresponding frequency, f_{source}, as follows.

$$(3\text{-}9) \qquad a_{grav} = \Delta v / T_{source} = \Delta v \cdot f_{source}$$

It can then be reasoned using equation 3-9 = equation 3-7 as follows.

$$(3\text{-}10) \qquad a_{grav} = \Delta v \cdot f_{source} = G \cdot \left[\frac{m_{source}}{d^2}\right]$$

Equation 3-11, below, is obtained by using that frequency is proportional to mass. With f_p and m_p as the proton frequency and mass then $f_{source} = [m_{source} / m_p] \cdot f_p$.

$$(3\text{-}11) \qquad \Delta v \cdot \left[\frac{m_{source}}{m_p}\right] \cdot f_p = G \cdot \left[\frac{m_{source}}{d^2}\right]$$

Rearranging and canceling m_{source} on both sides of the equation,

$$(3\text{-}12) \qquad \Delta v = \frac{G \cdot m_p}{d^2 \cdot f_p} \quad \text{per cycle of } f_{source}.$$

Then substituting, per equation 3-8, $m_p = [h \cdot f_p] / c^2$,

$$(3\text{-}13) \qquad \Delta v = \left[\frac{G}{d^2 \cdot f_p}\right] \cdot \left[\frac{h \cdot f_p}{c^2}\right]$$

$$= \frac{G \cdot h}{d^2 \cdot c^2} \quad \text{per cycle of } f_{source}.$$

The Planck Length, l_P, is defined as

$$(3\text{-}14) \quad l_P \equiv \left[\frac{h \cdot G}{2\pi \cdot c^3}\right]^{\frac{1}{2}} \quad \text{so that} \quad G = \left[\frac{2\pi \cdot c^3 \cdot l_P^2}{h}\right]$$

Substituting G as a function of the Planck Length from equation 3-14 into G as it is in equation 3-13, the following is obtained.

$$(3\text{-}15) \quad \Delta v = \left[\frac{2\pi \cdot c^3 \cdot l_P^2}{h}\right] \cdot \left[\frac{h}{d^2 \cdot c^2}\right]$$

$$= c \cdot \frac{2\pi \cdot l_P^2}{d^2} \quad \text{per cycle of } f_{source}.$$

This result states that:
- the velocity change due to gravitation, Δv,
- per cycle of the attracting mass's equivalent frequency, f_{source},
    which quantity, $\Delta v \cdot f_{source}$, is the gravitational acceleration, a_{grav},
- is a specific fraction of the speed of light, c, namely the ratio of:
    - 2π times the Planck Length squared, $2\pi \cdot l_P^2$, to
    - the squared separation distance of the masses, d^2.

That squared ratio is, of course, the usual inverse square behavior.

This also means that at distance $d = \sqrt{2\pi} \cdot l_P$ from the center of the source, attracting mass, the acceleration, Δv, per cycle of that attracting mass's equivalent frequency, f_{source}, is equal to the full speed of light, c, the most that it is possible to be. In other words, at that [quite close] distance from the source mass the maximum possible gravitational acceleration occurs. That is the significance, the physical meaning, of l_P or, rather, of $\sqrt{2\pi} \cdot l_P$.

The physical significance of $\sqrt{2\pi} \cdot l_P$ is that it sets a limit on the minimum separation distance in gravitational interactions and it implies that a "core" of that radius is at the center of fundamental particles having rest mass. That is, equation 3-15 clearly implies that it is not possible for a particle having rest mass to be approached closer than that distance.

That physical significance of $\sqrt{2\pi} \cdot l_P$, is so fundamental to gravitation and apparently to particle structure, that it more truly represents a fundamental constant than does l_P. For those reasons that length should replace l_P as a fundamental constant of nature as follows.

(3-16)   The fundamental distance constant, δ

$$\delta^2 \equiv 2\pi \cdot l_P^2$$

$$\delta = 4.051{,}34 \times 10^{-35} \text{ meters}$$

Equation 3-15 then becomes equation 3-17.

$$(3\text{-}17) \quad \Delta v = c \cdot \frac{\delta^2}{d^2} \quad \text{per cycle of } f_{source}$$

a quite pure and precise statement of gravitation: that gravitation is a function of the speed of light, c, and the inverse square law, in the context of the oscillation frequency, f_{source}, corresponding to the attract**ing**, source body's mass.

It makes clear that an oscillation is an integral part of gravitation as should be the case because gravitation is an action between particles having mass, which are the just-developed *Spherical-Centers-of-Oscillation* products, equation *3-6*, of the "Big Bang".

The Particle Core's Propagated Outward Flow

Each gravitationally attract**ing** *Spherical-Center-of-Oscillation* must tell each gravitationally attract**ed** particle its "message": the direction from the attract**ed** particle to the attract**ing** one and the magnitude of the attract**ing** particle's gravitational attraction. That task is assigned by contemporary physics' theory to a *gravitational field*, a vector field that is an assignment of a direction of action and its magnitude to each point in a region of space.

However, that designation of the field, while facilitating the description of the action fails to explain the cause, the mechanism of the field and thus fails to explain or account for the action at issue. It also fails to account for the time delay due to the limitation of the speed of light that must exist between a change at the attract**ing** particle and its effect at the attract**ed** particle.

Something flowing is required, something flowing at the speed of light, continuously, carrying the direction and magnitude information, spherically outward, from every gravitating *Spherical-Center-of-Oscillation* to every other *Spherical-Center-of-Oscillation*.

Furthermore, the necessity for gravitation that an oscillation and its frequency are closely involved in the effect [Equations *3-15* and *3-17*] and therefore in what is communicated by the flow, means that the flow itself is oscillatory corresponding to and generated by its oscillatory source, the *Spherical-Center-of-Oscillation*.

For such a flow to persist there must be a supply of that outward flowing substance in every particle. And, for that flow to have persisted the billions of years since the "Big Bang" that "supply" must be an extremely concentrated reservoir of that which flows outward [concentrated relative to the outward flow].

Having now just determined:
- That δ sets a limit on the minimum separation distance in gravitational interactions and therefore that a "core" of that radius is at the center of fundamental particles, and
- That an extremely concentrated reservoir supply of that which is flowing outward is required at the center of all particles to support the billions of years of their outward flow;

Therefore:
- The reservoir is the spherical "core" of radius δ at the center of all particles;
- That it is impenetrable is because of its immense density concentration [billions of years worth of flow of the flow substance [*Medium*] in the minute ($\delta = 4.05134 \times 10^{-35}$ *meters* radius spherical core) of every particle having rest mass], and.

- The *Spherical-Center-of-Oscillation* is a spherical oscillation of that immensely concentrated flow substance, *Medium*.

Then, what "contains" that core's supply or why doesn't it all just quickly "slosh" out and be gone ? The answer is that it is trying to do just that, to "slosh" out, as hard as it can. It cannot help propagating outward because it has no container. But it can only propagate outward at the limiting rate determined by its surface area, $4 \cdot \pi \cdot \mathcal{S}^2$ and the fastest speed possible for flow, the speed of light, c. Thus is the *Propagated Outward Flow*.

The Speed of the Flow – The Speed of Light

Every oscillation that we know in nature exhibits, and the very theory of oscillations in the abstract requires, that the oscillation consist of two aspects of the substance which is oscillating [e.g. pendulum position and velocity or electric potential and current] storing and exchanging back and forth the energy of the oscillation. With one aspect varying in oscillatory fashion then when that aspect decreases there must be some "place" for its energy to go, a place in which it is stored until it reappears in that aspect when it increases again. It cannot completely disappear or be lost because the oscillation would die. That "place" is the oscillation's second aspect and it obviously must vary in a manner related to the first aspect's variation with its energy storage in opposite phase.

The matter of the universe is largely a mass of particles each a spherical *[1 - Cosine]* form oscillation propagating outward.

Like electric inductance and capacitance determining the speed of propagation along a transmission line, μ_0 and ε_0 determine the speed of the *[1 - Cosine]* form oscillation propagation by setting the two aspects of the oscillation in which they are involved, the aspects between which the oscillation energy exchanges back and forth.

But, when the original oscillation came into existence it did so in absolute nothing. There was no "free space" with μ_0 and ε_0. There was nothing but the original oscillation. And, after the immediate explosion into all of the particles of the universe, each of those particles was sending its *Propagated Outward Flow* <u>into nothing, into emptiness</u>.

Where did the *Propagated Outward Flow*'s μ_0 and ε_0 come from? The only thing they could have come from was the original oscillation. There is no other possible source because everything else was absolute nothing, "the zero of existence". The μ_0 and ε_0 are inherent in the substance of the oscillation, which means, μ_0 and ε_0 are also inherent in the outward propagation. Each particle's *Propagated Outward Flow* contains its own μ_0 and ε_0.

Having established the supply of *Medium* [flow substance] and its on-going *Propagated Outward Flow* serving the role of gravitational field as a property of every particle exhibiting rest mass, the question arises, "What of the electric field, much stronger than gravitation and co-present with gravitational field whenever the gravitating particle has electric charge ?"

Just as is the case for gravitation, every particle having electric charge must tell its similar "message" to every other such particle. That requires something flowing outward at the speed of light continuously, carrying the direction and magnitude information, spherically outward, from every electrostatic *Spherical-Center-of-*

3 – THE FORM AND BEHAVIOR OF MATTER

Oscillation to every other *Spherical-Center-of-Oscillation*. That flow-communication is the electric field, an active process not a static state.

The theory of an *electric field*, just as with that of a *gravitational field*, above, while facilitating the description of the action fails to explain the cause, the mechanism of the field and thus fails to explain or account for the action at issue. It also fails to account for the time delay due to the limitation of the speed of light that must exist between a change at the attract**ing** particle and its effect at the attract**ed** particle

Two such simultaneous flows, gravitational and electric, and two supporting reservoirs supplying the flows, is clearly untenable. There can only be one reservoir in each particle's "core" and one resulting *Propagated Outward Flow* producing both the gravitational action and the electric action if for no other reason than because two supply reservoirs would mutually interfere with a spherically outward flow of each. There can only be one, single, universal flow

---

SUMMARY FOR – THE FORM AND BEHAVIOR OF MATTER

The form of matter is not that of the "particles" of classical modern physic's Standard Model. Rather the form of matter is:

- *Spherical-Centers-of-Oscillation*, spherical oscillations of *[1 - Cosine]* form, equation 3-6;

- Propagating spherically outward a continuous oscillatory *Propagated Outward Flow* of *Medium* in *[1 - Cosine]* form, according to its source *Spherical-Center-of-Oscillation* magnitude, sign, and frequency;

- The speed of the *Propagated Outward Flow*, c, being set by the net μ and ε in the *Medium* being propagated;

(3-18)
$$c = \frac{1}{\sqrt{\mu \cdot \varepsilon}}$$

The *Spherical-Center-of-Oscillation* consists of a central "core", a spherical volume of radius $\delta = 4.051,34 \times 10^{-35}$ meters that consists entirely of a high density concentration of the oscillating *Medium*, which propagates outward at an extremely low rate determined by being through the minute surface area of the "core" and the radial outward speed of flow of the propagated *Medium*, the speed of light, c.

Next: The motion of matter – the problem of relativity.

SECTION 4

Motion and Relativity

THE PROBLEM

A *Spherical-Center-of-Oscillation* naturally sends a *Propagated Outward Flow* of *Medium* uniformly radially outward in all directions from itself at velocity c, the speed of light, as presented in Section 3. The speed of that flow is set by the μ_0 and ε_0 of the *Medium* to the exact value of c by virtue of their controlling the cyclical alternating exchange of the oscillation between the two forms in which it exists.

When the center is not in motion that presents no problem, but with the *Spherical-Center-of-Oscillation* moving in some direction the center's motion and its propagation are in conflict. In the direction of motion the velocity of the center, v, tends to add to the natural value of the speed, c, of propagation of the *Propagated Outward Flow* and in the opposite direction it tends to subtract. But, the speed of the flow is fixed; set at c by μ_0 and ε_0.

That conflict forces an adjustment of the oscillation of the *Spherical-Center-of-Oscillation* to modify its propagation speed of its *Propagated Outward Flow*.

THE *Spherical-Center-of-Oscillation* at Constant Velocity

The treatment is of the *Spherical-Center-of-Oscillation* at constant velocity because that is the more direct and simple case of motion, and at constant velocity one cannot detect absolute motion. That is, one can say that there is a relative difference of velocity between two systems at constant velocity in one of which the observer is located, but the observer cannot say which system is moving and which, if any, is at rest.

To describe the behavior of the center its propagation will be modeled resolved into three components: forward, rearward, and sideward relative to the direction of the center's velocity, as depicted in Figure 4-1. [In the figure the "up", "down", "left" and "right" are all "sideward".] These orthogonal components represent the propagated wave in all directions. The wave in any particular direction is the "resultant" of that directions' projection on the forward or rearward component (whichever is at a nearer angle) and on the sideward component. (The "resultant" is the hypotenuse of the right triangle having the projection components as its other two sides.)

When a center is at rest [absolute "rest" relative to its propagation] then propagation of waves is the same in all directions at speed $c = \lambda_r \cdot f_r$.

A Center-of-Oscillation at Rest

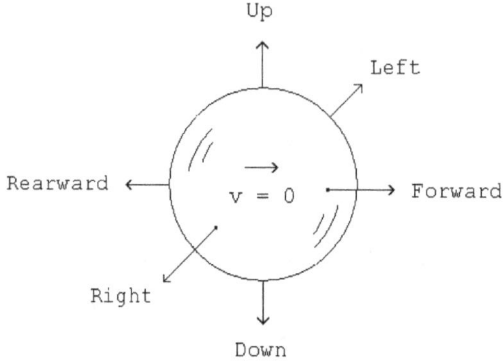

Figure 4-1

As described above under "The Problem" the speed of flow of centers' propagation is fixed at c by the μ_0 and ε_0 of the flowing *Medium*. The center moving at velocity v would find (in the forward direction) its freshly emitted propagation "thrown" forward at speed $[c + v]$ interfering with the flow just ahead of it at speed c and conflicting with the μ_0 and ε_0 of the *Medium*. It finds the propagated wave not moving out of the way at the needed $[c + v]$ in time for the next cycle as set by the at-rest frequency of the center. The result is an imperative to reduce the center frequency ["delay" the next cycle] by the factor $[1 - v/c]$. That "interfering" and "conflicting" tends to force on the center a change in its oscillation, a reduction by the factor $[1 - v/c]$. That is, with the center moving forward at v,

(4-1) Propagated Speed would become $c \cdot [1 - v/c] = (c - v)$

Flow speed = propagated speed + v = (c - v) + v = c

In the rearward direction the opposite is the case, an imperative to increase the center frequency by the factor $[1 + v/c]$. But, the *Spherical-Center-of-Oscillation* can only oscillate at one specific frequency at a time. It cannot both increase and decrease its oscillation frequency at the same time. It responds by adopting a compromise change in frequency, the geometric mean of the two conflicting factors as in equation *4-2*.

The center's oscillation frequency decreases and its oscillation wavelength correspondingly increases, the product still being c.

(4-2)
$$f_v = f_r \cdot \left[1 - \frac{v^2}{c^2}\right]^{1/2} \quad \text{[Center frequency decreases]}$$

$$\lambda_v = \lambda_r \cdot \frac{1}{\left[1 - \frac{v^2}{c^2}\right]^{1/2}} \quad \text{[Center wavelength increases]}$$

$$\lambda_v \cdot f_v = \lambda_r \cdot f_r = c \quad \text{[Wave velocity still at } c\text{]}$$

While the center can oscillate at only one frequency, it can propagate at different wavelengths in different directions. To maintain propagated wave velocity at c in the direction of center motion the wave must be actually propagated forward by the center at

$c-v$ relative to the center itself so that the wave velocity relative to at rest is the propagated velocity, c, plus the center velocity, v, that is $(c-v)+v = c$.

To propagate forward at $[c-v]$ while maintaining the frequency at f_v requires that the wavelength change to a smaller value, λ_{fwd}. Likewise, rearward the wave must be actually propagated by the center at $[c+v]$ relative to the center with a greater wavelength, λ_{rwd}. Those adjusted propagation wavelengths are as follows.

(4-3)
$$\lambda_{fwd} = \frac{c-v}{f_v} = \frac{c\left[1-\dfrac{v}{c}\right]}{f_r\left[1-\dfrac{v^2}{c^2}\right]^{\frac{1}{2}}} = \lambda_r \cdot \frac{\left[1-\dfrac{v}{c}\right]^{\frac{1}{2}}}{\left[1+\dfrac{v}{c}\right]^{\frac{1}{2}}} = \lambda_r \cdot \left[\frac{c-v}{c+v}\right]^{\frac{1}{2}}$$

$$f_{fwd} = \frac{c}{\lambda_{fwd}} = f_r \cdot \left[\frac{c+v}{c-v}\right]^{\frac{1}{2}}$$

$$\lambda_{rwd} = \lambda_r \cdot \left[\frac{c+v}{c-v}\right]^{\frac{1}{2}}$$

$$f_{rwd} = f_r \cdot \left[\frac{c-v}{c+v}\right]^{\frac{1}{2}}$$

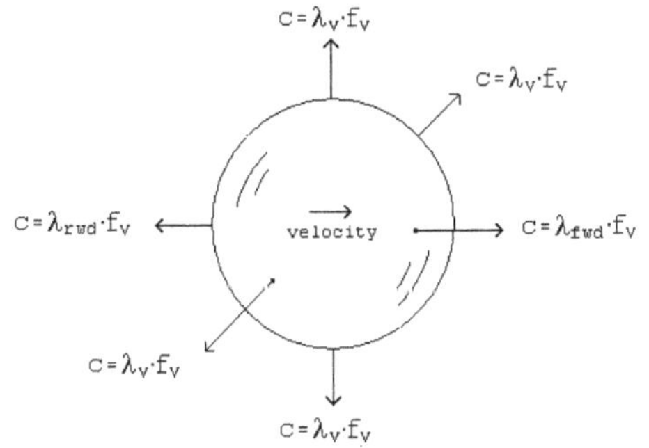

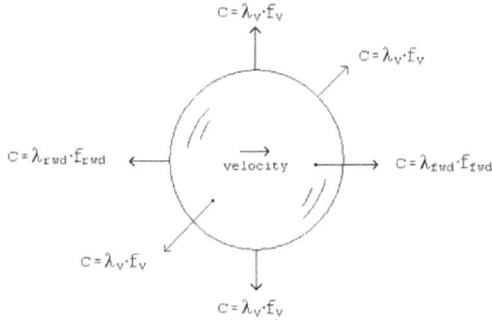

The Wave as Propagated by the Center at Velocity v (relative to the center)
Figure 4-2

The Above Propagation (as Observed from At-Rest)
Figure 4-3

As the center "sees" it, per the above Figure 4-2, it is oscillating at f_v, with the forward and rearward wavelengths adjusted for the velocity so that the wave travels in each direction at speed **c**. As "at-rest" would "see" it, per Figure 4-3, below, the center appears to propagate different forward and rearward frequencies, f_{fwd} and f_{rwd}.

Thus the field of propagated waves is traveling at c in all directions as observed by the center that is in motion and doing the propagating and as observed from at-rest.

What is "at rest"? It is the environment of a center not in motion.

THE EFFECT OF VELOCITY ON MASS

Of the total wave traveling outward from the source *Spherical-Center-of-Oscillation*, the only part that interacts with another, encountered, *Spherical-Center-of-Oscillation* is the part intercepted by the encountered center. The *Spherical-Center-of-Oscillation* intercepting the larger portion of incoming wave receives the greater impulse, the greater momentum change. Thus center mass depends on the encountered center's cross-section target for interception of *Propagated Outward Flow* waves. A *Spherical-Center-of-Oscillation* of smaller cross-section is of greater mass.

With the oscillation frequency corresponding to the rest mass of the particle it represents per Equations *3-6*, the development so far of decreasing oscillation frequency, equation *4-2*, demonstrates a decrease in rest mass due to the *Spherical-Center-of-Oscillation's* velocity. That is more properly referred to as a decrease in that part of the mass effect due to the overall frequency of oscillation of the center, to be referred to as "mass in rest form", m_r' in equation *4-4*.

(4-4)
$$m'_r = m_r \cdot \frac{f_v}{f_r} = m_r \left[1 - \frac{v^2}{c^2}\right]^{\frac{1}{2}}$$

However, overall the total mass increases because the effects so far have reduced the cross-section target for interception of *Propagated Outward Flow*.

From the forward or the rearward point of view the center's cross-section is the area of the circle of radius λ_v, the sideward direction. Per equation *4-2*.

(4-5)
$$\lambda_v = \lambda_r \cdot \frac{1}{\left[1 - \frac{v^2}{c^2}\right]^{\frac{1}{2}}}$$

Relative to the center's rest mass, m_r, the overall mass at velocity, m_v, is

(4-6)
$$m_v = m'_r \left[\frac{\lambda_v}{\lambda_r}\right]^2 = m_r \left[1 - \frac{v^2}{c^2}\right]^{\frac{1}{2}} \cdot \left[\frac{1}{\left[1 - \frac{v^2}{c^2}\right]^{\frac{1}{2}}}\right]^2$$

$$m_v = m_r \cdot \frac{1}{\left[1 - \frac{v^2}{c^2}\right]^{\frac{1}{2}}}$$

4 – MOTION AND RELATIVITY

From the sideward point of view the cross-section is no longer a circle, however. In the forward direction the at-rest circle's radius has become λ_{fwd} instead of λ_v and in the rearward direction λ_{rwd} instead of λ_v.

(4-7)
$$\lambda_{fwd} = \frac{c-v}{f_v} = \frac{c\left[1-\frac{v}{c}\right]}{f_v} = \frac{\left[1-\frac{v}{c}\right]}{\lambda_v} \quad \text{therefore} \quad \frac{\lambda_{fwd}}{\lambda_v} = \left[1-\frac{v}{c}\right]$$

(4-8)
$$\lambda_{rwd} = \frac{c+v}{f_v} = \frac{c\left[1+\frac{v}{c}\right]}{f_v} = \frac{\left[1+\frac{v}{c}\right]}{\lambda_v} \quad \text{therefore} \quad \frac{\lambda_{rwd}}{\lambda_v} = \left[1+\frac{v}{c}\right]$$

The product of the change factors, Equations 4-7 and 4-8, is $[1 - v^2/c^2]$, a reduction of cross-section, the same amount of increase in mass as equation 4-6.

THE LORENTZ CONTRACTIONS, LENGTH AND TIME

Logic requires of the overall universe that in all frames of reference at constant velocities relative to each other [*i.e.* inertial frames]:

- The equations describing the laws of physics have the same form, and

- The universal constants appearing in those equations be the same,

This is called the Principle of Invariance, and means that the speed of light, *c*, a universal constant, is the same in all inertial frames, which appears to conflict with our instinctive assumption that the speed of light should vary with the speed of the light's source.

That logic combined with experiments showing that the speed of light actually is the same independent of whatever inertial frame, required the development of the Lorentz Transformations to account for the constancy of the speed of light. The transformations are coordinate transformations between two inertial frames. The Lorentz contractions are the related change in the fundamental quantities: mass, length, and time.

The Lorentz Contractions

The Lorentz Contractions are as follows.

(4-9)
$$L = L_r \cdot \left[1 - \frac{v^2}{c^2}\right]^{\frac{1}{2}} \qquad \text{[Observed Length in the Direction of motion shortens.]}$$

$$f = f_r \cdot \left[1 - \frac{v^2}{c^2}\right]^{\frac{1}{2}} \qquad \text{[Observed frequency slows.]}$$

$$t = t_r \cdot \frac{1}{\left[1 - \frac{v^2}{c^2}\right]^{\frac{1}{2}}} \qquad \text{[Observed time periods length, Time passes more slowly.]}$$

(4-9 continued)

$$m = m_r \cdot \frac{1}{\left[1 - \frac{v^2}{c^2}\right]^{\frac{1}{2}}}$$ [Observed mass increases.]

Time and frequency are reciprocals of each other and the above equation 4-2 decrease in center frequency with velocity validates the f and t Lorentz Transforms. [The increasing λ_r to λ_v of that equation is compensating for the frequency decrease to keep the sideward propagation speed at c. Sideward is not the direction of v so the Lorentz Contraction does not apply to that λ.]

The equation 4-6 overall increase in center mass with velocity validates the mass, m, Lorentz Transform. Remaining to be validated is the length, L contraction. The λ_{fwd} and λ_{rwd} contraction Equations 4-7 and 4-8 are a center length contraction in the velocity direction, a Lorentz Contraction.

On the macroscopic scale it is necessary to investigate two centers and the distance between them in order to develop a velocity-caused contraction of length in matter. In bulk matter composed of multiple particles, atoms and their components, the spacing of the atoms depends on the balance of the various electrostatic forces acting as a result of the centers-of-oscillation, protons and electrons, of which the matter atoms are composed.

Considering just two *Spherical-Centers-of-Oscillation* at rest in a fixed position relative to each other, the effect of their moving jointly at velocity v in the direction of the line joining them should be a Lorentz Contraction to closer spacing of the two centers by the Lorentz Contraction factor.

The position of each of the two centers is the balance of all of the forces acting on the centers, an equilibrium position. If the velocity is to change the distance between the two centers then the force acting between the two centers must change so that a new closer equilibrium spacing exists and determines the new distance between the two centers. For the centers to need to be closer in order to re-establish equilibrium the effective charge of each of the centers must be decreased by the velocity.

In other words, for the Coulomb force between the two centers

(3-21) $$F = \frac{Q_1 \cdot Q_2}{4\pi \cdot R^2}$$

to be unchanged even though R is reduced by the Lorentz Contraction by the factor

(4-10) $$\frac{R_{vel}}{R_{rst}} = \left[1 - \frac{v^2}{c^2}\right]^{\frac{1}{2}}$$

so that R^2 is changed by the factor

$$\frac{R^2_{vel}}{R^2_{rst}} = \left[1 - \frac{v^2}{c^2}\right]$$

then $Q_1 \cdot Q_2$ must be so reduced by the same factor as is R^2.

But, that is exactly the case. It has already been shown by equation *4-3* that the forward wave propagation speed is reduced by the factor $[1-v/c]$ to $c' = c-v$ and that the rearward wave propagation speed is analogously changed by the factor $[1+v/c]$ to $c'' = c+v$.

The charge, Q enters into Coulomb's Law as the amount of *Propagated Outward Flow* actually propagated [how many multiples of the fundamental basic charge] times its speed, c, [see Appendix C, "Derivation of Coulomb's Law", equation C-16 so that the charge, Q, of the trailing center "looking" forward is reduced by the reduction of its c to $c' = c - v$, a factor of $[1-v/c]$.

Similarly the charge, Q, of the leading center "looking" backward is increased by the increase of its c to $c'' = c + v$, a factor of $[1+v/c]$.

Therefore, $Q_1 \cdot Q_2$ is modified by the product of the two factors which is $[1 - v^2/c^2]$, which matches the Lorentz Contraction of R^2 and therefore of R and validates the length, L, Lorentz Contraction.

The Velocity-Caused Impulse Increment

There is, however, another component to the interaction. While, in the forward direction, the source *Spherical-Center-of-Oscillation* propagates the wave at $c' = c-v$, the wave actually travels at velocity c because the center itself is traveling forward at v yielding the overall wave velocity as $c'+v = (c-v)+v = c$. The forward wave, the force it can deliver reduced by its propagation at $c-v$, is thus also "thrown forward" by the center's velocity. This adds another component of force, of potential impulse per wave times the wave repetition rate, the force that the wave can deliver to an encountered center.

In fact, without the wave having that additional component of force, and the consequent reaction back on the center in that same additional amount, the center would not experience equal reaction back on it in all directions from its propagated wave. The magnitude of this "force component" due to the center's velocity is $[v/c] \cdot F_r$, where F_r is the force that the wave would deliver if at rest and which it does always deliver to the sides: up, down, right and left.

The situation is analogous for the rearward wave otherwise the reaction back on the center by the rearward propagated wave would be $[v/c] \cdot F_r$ greater than the rest case. Without these "force components" the center would be self-accelerated in the forward direction by a force of $2 \cdot [v/c] \cdot F_r$ (the forward and rearward effects combined), clearly not the case.

Returning to the case of two *Spherical-Centers-of-Oscillation* traveling in the direction of an imaginary line joining them, when the forward wave of the trailing center encounters the rear of the leading center the $+[v/c] \cdot F_r$ positive "force component" of the forward wave and the $-[v/c] \cdot F_r$ negative "force component" of the rear of the encountered leading center cancel out leaving the net action due to the encounter as presented above before considering the "force component due to center velocity or momentum" aspect.

Particle Kinetics

The situation is the same with the rearward propagated wave of the leading center encountering the front of the trailing center. The net effect on the interaction is null, but the phenomena are still there.

Kinetic energy, KE, is defined as the work done by the force, f, acting on the particle or object of mass, m, over the distance that the force acts, s. This quantity is calculated by integrating the action over differential distances. It was done using the Lorentz Contraction for mass originally by Einstein as follows

$$(4\text{-}10) \quad KE = \int_0^s f \cdot ds \qquad \text{[Per above definition]}$$

$$= \int_0^s \frac{d(m \cdot v)}{dt} \cdot ds \qquad \text{[Newton's 2}^{nd}\text{ law,]} \quad [f = m \cdot a = m \cdot dv/dt]$$

$$= \int_0^{(m \cdot v)} \frac{ds}{dt} \cdot d(m \cdot v) \qquad \text{[Rearrangement of form]}$$

$$= \int_0^{(m \cdot v)} v \cdot d(m \cdot v) \qquad [v = ds/dt]$$

But, now the mass, m, increases with velocity per equation 4-9, Therefore:

$$KE = \int_0^v v \cdot d\left[\frac{m_r \cdot v}{\left[1 - \frac{v^2}{c^2}\right]^{1/2}}\right] \qquad \begin{array}{l}\text{[m is } m_r \text{ Lorentz} \\ \text{contracted by v.} \\ m_r \text{ is rest mass]}\end{array}$$

$$= \frac{m_r \cdot v^2}{\left[1 - \frac{v^2}{c^2}\right]^{1/2}} - m_r \cdot \int_0^v \frac{v \cdot dv}{\left[1 - \frac{v^2}{c^2}\right]^{1/2}} \qquad \text{[Integration by parts]}$$

$$(4\text{-}11) \quad KE = \frac{m_r \cdot v^2}{\left[1 - \frac{v^2}{c^2}\right]^{1/2}} + m_r \cdot c^2 \cdot \left[1 - \frac{v^2}{c^2}\right]^{1/2} - m_r \cdot c^2 \qquad \text{[Integration of 2nd term]}$$

$$= \frac{m_r \cdot v^2 + m_r \cdot c^2 \cdot \left[1 - \frac{v^2}{c^2}\right]}{\left[1 - \frac{v^2}{c^2}\right]^{\frac{1}{2}}} - m_r \cdot c^2 \quad \text{[Place 2}^{\text{nd}}\text{ term over 1}^{\text{st}}\text{ term denominator]}$$

$$= \frac{m_r \cdot v^2 + m_r \cdot c^2 - m_r \cdot v^2}{\left[1 - \frac{v^2}{c^2}\right]^{\frac{1}{2}}} - m_r \cdot c^2 \quad \text{[Expand term with brackets]}$$

$$KE = \frac{m_r \cdot c^2}{\left[1 - \frac{v^2}{c^2}\right]^{\frac{1}{2}}} - m_r \cdot c^2 \quad \text{[Simplify]}$$

$$= m_v \cdot c^2 - m_r \cdot c^2 \quad \text{[}m_v\text{ is total mass at }v > 0$$
$$m_r\text{ is total mass at }v = 0$$
$$m_v = m_r \text{ Lorentz transformed]}$$

This result states that:

{Kinetic Energy} = {Total Energy} - {Rest Energy}

or

{Total Energy} = {Kinetic Energy} + {Rest Energy}

    The appearance in this result that the energies are the product of the masses times c^2, the speed of light squared, was the origination of that concept, the famous Einstein's $E = m \cdot c^2$. The concept falls out naturally from applying the Lorentz transforms to the classical definition of kinetic energy. It is somewhat surprising that Einstein was the first to do that inasmuch as it was Lorentz who developed the Lorentz transforms and the Lorentz contractions.

Alternative Treatment of the Same Derivation

    If in the above original derivation one proceeds differently from the first line of equation *4-11* on, as below, a slightly different result is obtained.

(4-11)
$$KE = \frac{m_r \cdot v^2}{\left[1 - \frac{v^2}{c^2}\right]^{\frac{1}{2}}} + m_r \cdot c^2 \cdot \left[1 - \frac{v^2}{c^2}\right]^{\frac{1}{2}} - m_r \cdot c^2 \quad \text{[Repeated (4-11) to start here]}$$

(4-12)
$$KE + m_r \cdot c^2 = \frac{m_r \cdot v^2}{\left[1 - \frac{v^2}{c^2}\right]^{\frac{1}{2}}} + m_r \cdot c^2 \cdot \left[1 - \frac{v^2}{c^2}\right]^{\frac{1}{2}} \quad \text{[Move the right most "} - m_r \cdot c^2 \text{"]}$$

Considering and evaluating the three terms of equation *4-12*:

$KE + m_r \cdot c^2$ = Kinetic plus rest energies
= Total Energy
= $m_v \cdot c^2$

$\dfrac{m_r \cdot v^2}{\left[1 - \dfrac{v^2}{c^2}\right]^{1/2}}$ = A relativistically increased energy of motion.
= $m_v \cdot v^2$

$m_r \cdot c^2 \cdot \left[1 - \dfrac{v^2}{c^2}\right]^{1/2}$ = A relativistically reduced rest energy.
= $m_v \cdot c^2 - m_v \cdot v^2$

the result is that equation *4-12* is equivalent to

(4-13) $\begin{bmatrix}\text{Total}\\\text{Energy}\end{bmatrix} = \begin{bmatrix}\text{Energy in}\\\text{Kinetic Form}\end{bmatrix} + \begin{bmatrix}\text{Energy in}\\\text{Rest Form}\end{bmatrix}$

$m_v \cdot c^2 = m_v \cdot v^2 + m_v \cdot (c^2 - v^2)$

and (dividing the above energy equation by c^2 to obtain an equation in mass)

(4-14) $\begin{bmatrix}\text{Total}\\\text{Mass}\end{bmatrix} = \begin{bmatrix}\text{Mass in}\\\text{Kinetic Form}\end{bmatrix} + \begin{bmatrix}\text{Mass in}\\\text{Rest Form}\end{bmatrix}$

$m_v = m_v \cdot v^2 / c^2 + m_v \cdot (1 - v^2/c^2)$

The m'_r "mass in rest form", of equation *4-4*

$$m'_r = m_r \cdot \left[1 - \dfrac{v^2}{c^2}\right]^{\tfrac{1}{2}}$$

equals the m_v of equation *4-6*, below   multiplied by the $(1 - v^2/c^2)$ as in the bottom of equation *4-14*

$$m_v = m_r \cdot \dfrac{1}{\left[1 - \dfrac{v^2}{c^2}\right]^{\tfrac{1}{2}}} \qquad \left[1 - \dfrac{v^2}{c^2}\right]$$

Why is the formulation for classical *Kinetic Energy* $KE = \tfrac{1}{2} \cdot m \cdot v^2$ but *Energy in Kinetic Form* is simply $m \cdot v^2$ without the ½? When dealing with quite small velocities (v very small relative to c) the excursion of total energy above rest energy and the excursion of energy in rest form below rest energy are both essentially linear. In that case the portion above the rest case is essentially half of the total excursion above and below the rest case. The classical kinetic energy is then half, $\tfrac{1}{2} \cdot m \cdot v^2$, the total energy in kinetic form, $m \cdot v^2$, for $[v/c]$ quite small.

The Center of Oscillation "At Rest" and "In Motion"

In motion at a constant velocity, v, the *Spherical-Center-of-Oscillation* experiences the asymmetrical distortions of equation *4-3* and Figures 4-2 and 4-3. The distortions indicate the motion and the motion enhanced energy of the center. At rest, in the absence of motion the center is spherically symmetrical.

Thus the rest mass and rest energy correspond to the spherically symmetrical portion of the center's oscillation [the only portion if $v = 0$] and they are "mass in rest form" and "energy in rest form". The overall distorted portion corresponds to the total "mass in kinetic form" and "energy in kinetic form" of the center. Of course the difference of the two is the mass and energy in kinetic form.

> *- What, then, is the prime frame of reference ?*
>> *It is the "rest frame".*
>
> *- And, what is the "rest frame" ?*
>> *It is the frame in which particles are at rest.*
>
> *- And, how do particles exhibit their being at rest ?*
>> *By their oscillation and their Propagated Outward Flow*
>> *both being perfectly the same in all directions, that is*
>> *spherically symmetrical.*

SECTION 5

Prime Objective Space

THE PROBLEM

"What is motion, motion relative to what?" After all, the Earth and anything on its surface rotate about the Earth's axis, revolve around the sun, participate in the sun's motion in the galaxy and in the galaxy's motion through space. Thus use on the Earth's surface of the terms "static" or "in motion" requires clarification.

This is the fundamental problem underlying relativity, and it became a major issue upon the development of physics' treatment of electro-magnetic waves: is there a medium in which the electro-magnetic waves exist, and if so is it a "stationary" all-pervasive "aether", a prime reference system to which everything else is relative ? If not, what is the meaning of "static" or "in motion" and what of the motion of things relative to each other ?

The problem and its significance can be further appreciated by means of an example. We take a straight wire in which positive charge flows at constant velocity [constant speed and direction along the wire relative to the wire]. Classically, in terms of magnetic field behavior, there is a magnetic field circumferential to the wire. This field will exert a force on a charge moving in the field. Now, we, the observers, take on a velocity identical to the charge moving in the wire, the charge causing the magnetic field. In this case, to us, the charge in the wire is static. It is not moving and there should be no field. [It is true that to us in this case the wire appears to be traveling "rearward", but moving wires are not, in themselves, a cause of magnetic field.] Is there, now, as we view it, a magnetic field ? That is, from the "static", as we view it, charge ?

How do we reconcile this: a charge ""at rest"" relative to the Earth exhibits to us only static effects even though moving through space at a speed of at least 66,600 miles per hour [the Earth's speed around the sun] and a charge "at rest" relative to us [the above example of the wire] exhibits magnetic effects ?

RELATIVITY AND INVARIANCE

By the time of Newton and the development of his laws of motion it was well understood that all motion is relative to some frame of reference. One cannot say that something is moving at a stated velocity except by defining what the velocity is relative to. Newtonian mechanics dealt with this problem, successfully for "Newtonian systems". Direct linear relationships transfer Newtonian motion descriptions from one frame of reference to another.

In the second half of the 19th century Maxwell developed his equations describing electro-magnetic field, the equations being an outgrowth of the then developing understanding of electricity, charge, magnetic effects, and so forth. Substantially before the first actual detection of electro-magnetic waves by Herz toward the end of the century, it was recognized that Maxwell's equations described a wave propagating in space at a velocity, c, determined by two constants in the equations, ε and μ, the dielectric constant and the permeability of whatever medium the waves were passing through, such that $c^2 = 1/\varepsilon \cdot \mu$.

This result presented two problems.

First

At the time it seemed inconceivable that these [or any] waves could propagate other than in some medium. Since the waves could and do propagate throughout free space as well as through the air and through other substances some kind of all-pervading medium, called in those days an "aether", was postulated.

Second

Maxwell's equations would not correctly transform from one frame of reference to another at different velocity using the Newtonian transformations. Therefore it was thought that Maxwell's equations applied only to one, prime, frame of reference, that of the "aether", which also defined μ, ε, and, therefore, c.

[The Newtonian transform between two systems at different velocities is to merely subtract the velocity difference. For example, to a passenger in a train going forward at 30 miles per hour the train is a stationary reference system and the landscape out the window is traveling backwards at 30 miles per hour. To do a Newtonian transform from the train-as-reference to the landscape-as-reference one subtracts the landscape's 30 miles per hour backward from the landscape (making it stationary) and also from the train (making it to be going 30 miles per hour forward).

[If one attempts such a Newtonian transform on Maxwell's equations and the speed of light wrong results are obtained because of non-linearity. In addition, one cannot subtract a velocity difference between two systems from the speed of light, c, because c is an absolute constant given by $c^2 = 1/\varepsilon \cdot \mu$ and cannot vary with some other velocity.]

The problem in the assumption that there is an "aether" which is the electro-magnetic wave medium is that all attempts to define and detect the "aether" led to contradictions or further problems. The most famous of those attempts was the Michaelson-Moreley experiment, which, expecting to find two different measured results for the speed of light because of the motion of the earth in its orbit relative to the "aether", obtained the "negative" result that the speed of light always measured to be the same regardless of the motion of the observers, Michaelson and Morely and the Earth.

The Michaelson-Moreley experiment and the Newtonian transformation inadequacy required that a new transformation system be developed. That was done by Lorentz. Lorentz retained the existence of an "aether" which had to be the prime frame of reference. His transformations and their consequent "contractions" resolved the "aether" problems. The Lorentz transforms and the Lorentz contractions are familiar to all physicists and are fundamental to the Theory of Relativity.

In the early 1900's Einstein took the further step of denying that any "aether" or medium was necessary for electro-magnetic waves and that there was no prime frame of reference. Those assumptions were embodied in his Theory of Relativity for which, there being no "aether", everything is relative. The repeated failure to successfully define and detect an "aether", coupled with Einstein's formulation that dealt with the problem by denying the "aether's" existence, resulted in the complete acceptance of Einstein's theories and the abandonment of the "aether" problem. However, Einstein had no proof, only his opinion, to justify his aether denial.

Excepting only the issue of whether an "aether" exists and is the prime frame of reference, the Lorentz and the Einstein formulations are equally valid descriptions of physical reality. However, the Theory of Relativity and other developments in physics that came from Einstein [his explanation of the photoelectric effect and his famous $E = m \cdot c^2$] were tremendously successful. Relativistic effects could be observed and measured experimentally. The mass-energy equivalence was dramatically confirmed.

Just as Einstein had his doubts about some of the then accepted aspects of traditional 20th Century physics [in referring to some aspects of uncertainty and quantum mechanics he is reputed to have said that he "... did not believe that God plays with dice ...."] so Lorentz still clung to the necessity of an "aether" and the prime frame of reference that it implied.

But the relativity "bandwagon" was rolling and relativity carried the day. Perhaps that was, at least partly, because Lorentz died at 75 years old in 1928, a time when these issues were riling physics, and at that time Einstein was still young, not having died until 1955 at age 76.

New developments in space research long after the death of Lorentz and Einstein now make it necessary to reverse that outcome and conclusion. It can now be shown that Lorentz was essentially correct and Einstein incorrect with regard to a prime frame of reference and a medium in which electro-magnetic waves propagate. That is, there is a universal absolute frame of reference to which all motion is relative and it is the prime frame of reference. And, there is a medium, so to speak an "aether", it is the *Propagated Outward Flow* from all particles, penetrating everything, and filling the universe

It is now necessary to restate relativity more correctly. There is nothing inherent in Einstein's Theory of Relativity requiring his comprehensive relativity, the absence of a prime frame of reference. The concept "relative" does not necessarily enter into the mathematical derivations and "theory of relativity" is a misnomer. The theory-system called the Theory of Relativity should be correctly referred to as the "Principle of Invariance". Einstein's postulates were solely Invariance.

"Invariance" means that the laws of physics, the behavior of all physical reality, is the same in any and every coordinate system or frame of reference. Invariance requires that the form of the mathematical statements describing reality and the constants appearing in those statements be invariant under any transformation of coordinates, which means that they must be unchanged by any change of frame of reference regardless of its motion so long as it is at constant velocity with no acceleration involved. Since all universal constants appearing in equations describing physical reality are invariant, the speed of light, one of those constants, is invariant.

The principle of Invariance is not magical or mysterious, but obvious. When one walks down the street, breathes, throws a stone or rides in a space ship one is doing a thing. The thing is not changed by changing the frame of reference from which someone observes it. The act is invariant therefore its description must be so. Einstein's principal mistake was that while he recognized that Invariance was essential he did not look for a mechanism to cause that to be so, and the only possible such mechanism is a universe-wide single absolute frame of reference.

To be perfectly clear about this replacement of relativity with "absolutivity" the pertinent factors are as follows.

(a) All motion is absolute, that is, it is relative to an absolute, prime frame of reference.

> In normal human experience the absolute frame of reference cannot be detected so that motion seems to be relative, but that is only an appearance.

(b) The absolute frame of reference is not a "preferred" frame of reference in the sense of having special or different physical laws. It is a "prime" reference system in that all physical reality is relative to it.

> That is why the universe is invariant. For physical reality there is only one grand system of reference for everything. The universe does not "know" about our frames of reference; it simply is in its natural frame of reference, everywhere. It would be ridiculous for it not to be invariant.

This goes counter to some of the most basic accepted concepts of 20th Century physics. Consequently, it requires substantial justification, which is as follows.

(1) A medium is required for electro-magnetic waves. They either propagate in a medium or are themselves propagation of the wave "substance" or else they have no existence. Since they exist, and since their propagation is a transverse wave, not longitudinal, and since there has never been a contention that electro-magnetic waves involve motion of anything in the direction of wave propagation other than that of the wave's energy and momentum, the medium must exist.

One cannot say that there is no electro-magnetic wave medium just "field". "Field" is merely a code-word for "action at a distance", an inability to actually explain the mechanism and actions involved.

A medium is also required to define and set the propagation velocity of the waves to c, the speed of light. Without a medium there is no cause of a universal fixed value of c nor μ and ε, the dielectric constant and permeability of free space.

It develops that the medium of E-M propagation is the *Propagated Outward Flow* of all particles, the flow that carries the propagation's μ_0 and ε_0. See "The Medium", below.

(2) If one reasons in terms of Einstein's General Theory of Relativity, "curved" space-time, due to the variation of gravitation with the distribution of mass in the universe, and the gravitational field pervading the universe with its shape due to that variation, is itself a frame of reference. Since space-time is not uniformly "flat", the shape variations make possible detection not only of acceleration but also of absolute velocity relative to the total mass as distributed in the universe.

But, that reference frame is identical to the reference frame of the singularity at which the universe started with the "Big Bang".

(3) There exists throughout the universe a background radiation which is the residual radiation from the immense energy of the "Big Bang", the start of the universe. The temperature has now cooled down from the extremely high levels at the beginning to only about $2.7°$ *Kelvin*. That radiation is, of course, relative to the beginning, relative to "where the "Big Bang" took place. Measurements of Doppler frequency shift of this radiation due to the motion of the Earth give an absolute velocity for the Earth relative to the medium of about 370 *km/sec*. The absolute direction of the Earth's motion as indicated by those measurements is off in the direction from Earth of the constellation Leo.

That absolute velocity of the Earth is sufficiently low that observations from Earth are equivalent [within the accuracy involved] to observations from "at rest" in the absolute frame of reference.

(5-1) $$v_{Earth} \sim 370 \ km/sec$$

$$\left[1 - \frac{v_E^2}{c^2} \right]^{\frac{1}{2}} = 0.999,999,2 \ldots$$

(4) The Lorentz contractions must actually occur as shown in Section 4, not be mere observational effects. According to the theory of relativity, an object in motion experiences slower time. If two identical clocks agree and one clock is then moved away and returned while the other is motionless [in relativistic terminology if one is moved away and then returned relative to the other from which observations are made] the moved clock must read an earlier time than the unmoved clock even when both are again "at rest" in the same frame of reference. When both are so again together and "at rest" there can be no observational quirk to

cause them to read different times. The moved clock must have actually run slower.

It could be argued that the moved clock had to be accelerated to be moved so that the overall process was not a constant velocity situation. That is not the contention of relativity, however, which states that the moved clock does run slower and relies on the fact of acceleration to make the distinction as to which clock was moved and which stayed "at rest".

(5) Consider three clocks, #1, #2, and #3, at constant velocities v_1, v_2, and v_3. According to relativity the time of clock #3 is contracted by some amount relative to Clock #1. Likewise Clock #3 is time contracted relative to Clock #2, but by some different amount. But, Clock #3, with a time contraction relative to Clock #1 in an amount based on the velocity difference between Clock #1 and Clock #3, and with a time contraction relative to Clock #2 based on the velocity difference between Clock #2 and Clock #3, cannot be actually contracted two different amounts at the same moment. Since the contraction must be actual, not solely observational, an absurdity results.

The solution to this problem is simple. All clocks are actually, as observed from the prime frame of reference, contracted according to their absolute velocity relative to that frame, not according to their velocity relative to another moving clock. In addition, an observer at a moving clock observes somewhat different results than those actual absolute contractions because his standards of measurement have also been contracted by his own motion [even though they appear unchanged to him]. This produces an observed, but not actual modification of the absolute, actual contraction.

[Of course, if one of the moving clocks is moving at a modest velocity the difference between its "at rest" dimensions and its actual contracted ones is so small that the observations from that slow-moving clock would be essentially equivalent to from "at rest", the very case set out for planet Earth in (3) above.]

In his original paper on relativity Einstein contended that there was no way that an observer experiencing acceleration could distinguish between whether his system was actually accelerating in a region free from gravitation or was actually "at rest" in a gravitational field. In fact, that contention is incorrect and the distinction can be made by local measurement, as is now known. The distinction occurs because gravitation follows an inverse square law in practice in the real universe and gravitation is inherently radial relative to the gravitating mass.

One could say that Einstein was largely correct but for partially incorrect reasons. The same can be said of the effect of absolutivity on cosmology and space-time physics. The results obtained by traditional 20th Century physics and the theories leading to them are largely correct. Absolutivity only restores the medium and the prime [but not "preferred", special, nor having different physical laws] frame of reference.

The fact that until recently we could detect no absolute velocity and that even now it is only detectable with special scientific effort does not mean that all motion is relative, it only means that we have not developed the means for ready detection of absolutivity. There have been many other things that were undetectable in the past but that are not so now: germs, distant stars, x-rays, atoms, etc.

The Theory of Relativity has required mind-twisting adjustments to way of thinking, adjustments away from the reasonable and "apparent" to a mass of paradoxes and their proposed resolutions. Absolutivity retains contact with reality both in describing physical reality accurately and by doing so in a fashion much more consistent with reasonableness.

With absolutivity the principle of invariance becomes simple, practical and apparent in addition to being necessary as it always was. There is only one "system", the universe with some parts moving in various ways and some parts "at rest" and that one system has, of course, one overall set of physical laws throughout. Before absolutivity, invariance was necessary but was crying for an explanation, a cause. One can see no particular reason why invariance should be necessarily automatically true in the universe of the Theory of Relativity. Absolutivity solves the problem by showing the natural inevitability of invariance.

THE MEDIUM

Why does this new medium succeed when all prior attempts to define an "aether" without contradictions failed? The reason is the nature of the electro-magnetic wave medium, as follows.

Electro-magnetic field is cyclically changing electric and magnetic field. It is caused by changing motion of electric charge, of *Spherical Centers of Oscillation*. The changes are changes to the always-present static field of electric charge "at rest" which field is the "at rest" *Propagated Outward Flow*, see Appendix C, "Derivation of Coulomb's Law. The variations in the static field are changes in charge position, its location. The variations in the magnetic field are changes in velocity, further variation in the static field, a distortion of it due to the effect of charge velocity.

The electro-magnetic wave is a modulation of the *[1 - Cosine]* oscillation of the *Propagated Outward Flow*.

That propagation, the *Propagated Outward Flow* from all *Spherical Centers of Oscillation*, the static electric field, is the medium, the "aether" in which electro-magnetic field exists, and it is relative to the universe' prime frame of reference, that of the "Big Bang", that of where its source charges originated, where they were before motion carried them elsewhere. That propagation emanates from each charge, originally from the origin of the "Big Bang" and now from wherever each charge is. It, itself carries the controlling parameters μ and ε.

SOLUTION OF THE PROBLEM

It is now time to address the apparent paradox that was left as a question at the beginning of this discussion. The apparent paradox had two elements.

First

> A charge "at rest" relative to the Earth's surface exhibits to us, who are also "at rest" relative to the Earth's surface, no magnetic field even though the charge is clearly in motion with the Earth's surface rotating about the planet's axis, revolving about the sun and moving in and with the galaxy.

Second

    A charge in motion in an electric wire [as a current] does exhibit a magnetic field to us, who are [in this problem] moving with the same velocity as the charge even though the charge is "at rest" relative to us.

Although there are these two elements to the problem, they are one overall problem, an apparent inconsistency in physical laws. The inconsistency results directly from relativity and resolves when absolutivity is applied.

Considering first the problem of the wire, absolutivity answers with the solution,

    "Since the current in the wire is in absolute motion, it exhibits the usual magnetic field regardless of the motion of the observer. The only effect of the observer's motion is to change his standards of measurement and, therefore, the magnitude of the magnetic field as he measures it."

Relativity responds,

    "No, the explanation is that, although the current of the charge moving relative to the wire is zero relative to the observer moving at the same velocity, the overall wire including the charge is electrically neutral so that the wire moving 'rearward' without the charge [as the observer sees it] is an opposite charged wire moving in the opposite direction and produces the same magnetic field to the observer as he would see if he were "at rest" relative to the wire and he were observing the charge moving 'forward'. In other words, a wire moving 'rearward' while its current stands still gives the same magnetic field as the wire standing still and its current moving 'forward'."

Absolutivity then closes the discussion with,

    "If relativity were valid that would be a true and good analysis, but the same problem as that of the wire can be stated for a beam of charged particles in empty space without the wire. In such a case the magnetic field behavior is the same, the paradox for relativity is the same, but there is no 'wire' to travel 'rearward'. Thus, only the explanation of absolutivity will resolve the problem."

[This also illustrates the simplicity of absolutivity as compared to the complications of relativity.]

    The first part of the paradox, that of the charge "at rest" on the Earth's surface, is simply a case of magnitudes. In fact the charge "at rest" relative to the moving Earth is in absolute motion and does exhibit the expected magnetic field. However, the field is too small to be noticed. The magnitude of magnetic field is less than the corresponding electric field magnitude by a factor of $[v^2/c^2]$, the v being the velocity of the charges whose motion, as electric current, produces the magnetic field. The velocity of Earth [presented earlier above] is less than 10^{-3} of the speed of light so that $[v^2/c^2] < 10^{-6}$.

    In addition, of course, the Earth is overall electrically neutral and the magnetic field due to its motion in space consists largely of a pair of equal and opposite such fields.

5 – OBJECTIVE SPACE

SUMMARY

This brings us back to the point that, contrary to Einstein, there is an absolute frame of reference, an ""at rest"" frame. When the *Spherical-Center-of-Oscillation*'s oscillation is perfectly spherically symmetric, its propagation the same in all directions, then the center's velocity is zero and it is completely "at rest". That is the universe's absolute frame to which all motion and all other frames are relative.

That is why the Principle of Invariance is valid. That is why it is required of the overall universe that in all frames of reference at constant velocity relative to each other [*i.e.* inertial frames]:

- The equations describing the laws of physics have the same form, and

- The universal constants appearing in those equations are the same,

- The speed of light, c, a universal constant is the same everywhere.

They are all part of the same one overall absolute frame of reference, the "rest frame". The "rest frame" is not special in that its laws and constants are not different. It is special in that all other frames are relative to it. It is simply the actual frame of the "Big Bang".

> *- What, then, is the prime frame of reference, objective space?*
>
> *It is the "rest frame".*
>
> *- And, what is the "rest frame"?*
>
> *It is the frame in which particles are "at rest".*
>
> *- And, how do particles exhibit their being "at rest"?*
>
> *By their oscillation and their Propagated Outward Flow*
> *both being perfectly the same in all directions, that is*
> *spherically symmetrical.*

SECTION 6

The Effect of Prime Objective Space on Science

There are several areas or aspects of physics which must be treated completely differently in taking account of the prime frame and the consequent absolute space and absolute motion. These include:

[a] - the behavior of the orbital electrons of atoms in their effect on their line spectra and

[b] - cosmology and astronomy, in particular the behavior, history and future of the universe and the nature of space.

ATOMIC LINE SPECTRA

Fine Structure and Spin

When the line spectrum of Hydrogen is obtained with a spectrometer of high resolving power it is found that the lines that appear as simple single lines at low resolving power are in fact pairs of lines. This phenomenon is referred to as the "fine structure". The splitting of the (low resolution) single line into (high resolution) two lines is on the order of about 1 part in 104. Sommerfeld addressed this problem showing that if the orbital electrons had elliptical orbits, in which the electron velocity would be relatively slow far from the nucleus and faster than for the circular orbit case near the nucleus, the relativistic mass increase at the higher velocity provided a minute energy increase that was on the order of the correct amount to account for the line splitting. That is, the elliptical orbit's energy would be slightly different from a circular orbit's energy.

Sommerfeld's model for how the fine structure arises, a model based upon the conceived direct motion and action of the electrons, was soon superseded by Quantum Mechanics, a model that seeks not to directly represent electron motion but rather to express the electron behavior and its effects. However, in spite of the wide spread acceptance of Quantum Mechanics, the concept of elliptical electron orbits has been retained.

Quantum Mechanics overthrew the Bohr-Sommerfeld theory shortly after its development. In Quantum Mechanics the fine structure is attributed to the interaction of the magnetic field due to the electron's spin on its own axis with the magnetic field due to the electron's orbit around the nucleus. This is referred to as "spin-orbit coupling". The two cases that are contended to account for the two lines close together in the Hydrogen spectrum are for the electron's spin angular momentum vector in the same direction as the orbital motion angular momentum vector and in the opposite direction.

In multi-electron atoms the fine structure becomes various multiplet structures depending on the number of electrons, rather than the doublet structure of Hydrogen with only one electron. For multi-electron atoms the coupling possibilities are spin-orbit, orbit-orbit, and spin-spin.

In a sense the conception that traditional 20th Century physics has of the electron is of a powder of negatively charged minute specks compressed into a little ball. (One of the concerns of traditional 20th Century physics was that of what holds the electron together; with all of that charge packed so closely why does it not explode ?) In that sense, the electron is conceived of as spinning on its axis. It is conceived that the consequent circular motion of the specks of charge that are rotating about the electron's spin axis constitute a small current and generate a small magnetic field.

Actually, traditional 20th Century physics does not know, and has no way of knowing, whether the electron spins or not and if so then how rapidly, how (in traditional 20th Century physic's terms) the charge is distributed throughout the electron and what the electron diameter is, and so forth, all data necessary for calculation of its spin magnetic field. The contention of electron spin and its associated magnetic field depends entirely on that the concept is used to explain a fine characteristic in atomic line spectra. The amount of spin and the amount of consequent magnetic field is set by 20th Century physics at the value that explains the spectral fine structure.

In the present physics there can be no such concept, of course. Whether a *Center-of-Oscillation* can or does spin or not might conceivably be open to question but would seem to be inconsequential and irrelevant. The Coulomb effect of an electron *Center-of-Oscillation* is an effect external to the internal structure of the center. There is no way that such an electron can have a magnetic field due to spin.

Fine structure is the result of each orbital electron's having one or the other of two possible slightly different energy states in its orbit. In traditional 20th Century physics the two energy states result from the electron spin angular momentum (and magnetic) vector being in the same or opposite direction relative to the orbital motion angular momentum (and magnetic) vector. Spin in fact not being the cause because there is no spin, there must be some other cause that produces the same effect.

There is such another cause. That other cause is <u>absolute motion</u>, motion relative to the prime frame, the effects of which have been neglected until now in the treatment of the behavior of the atomic orbital electrons. Paraphrasing a portion of the earlier Section 4 - Motion and Relativity:

> "There exists throughout the universe a background radiation which is the residual radiation from the immense energy of the "big bang", the start of the universe. ... This radiation is, of course, relative to the beginning, relative to the U-wave medium. Measurements of Doppler frequency shift of this radiation due to the motion of the Earth give an absolute velocity for the Earth

relative to the medium of about 370 km/sec. The direction of the Earth's motion as indicated by those measurements is off in the direction from Earth of the constellation Leo."

The speed of the Earth in its orbit around the Sun is only about 31,000 m/sec so most of Earth's absolute speed is due to its motion relative to its galaxy, the Milky Way, and the absolute motion of that galaxy through space. Generally speaking it is likely that most if not all of the universe has a comparable magnitude of absolute velocity directed radially outward from the location of the original "big bang". (This is treated further in Appendix E, "The 'Big Bang' Outward Cosmic Expansion") But, whether or not, this absolute velocity of our Earth and our entire planetary-solar-galactic system of about $3.70 \cdot 10^5$ m/sec = $0.0014 \cdot c$ must be taken into account in considering the behavior of the orbital electrons.

The most important factor in the stability of an atomic orbital electron is that it must not radiate energy. That requires that it experience no changes in the shape of its *Propagated Outward Flow* pattern of propagation forward, rearward and sideward. And, that requires that its speed remain constant. But, the speed of an orbital electron has two components: its orbital speed relative to the nucleus and its absolute linear speed because it is part of our overall solar system motion in space.

In order for the electron to avoid radiating, it is its net speed, the resultant of those components, which must remain constant. The way in which those two components combine to produce a net electron speed at any moment depends upon the orientation of the electron's orbital plane relative to the absolute velocity component of the electron, its atom and its solar-galactic system. The effect is illustrated in Figure 6-1, below.

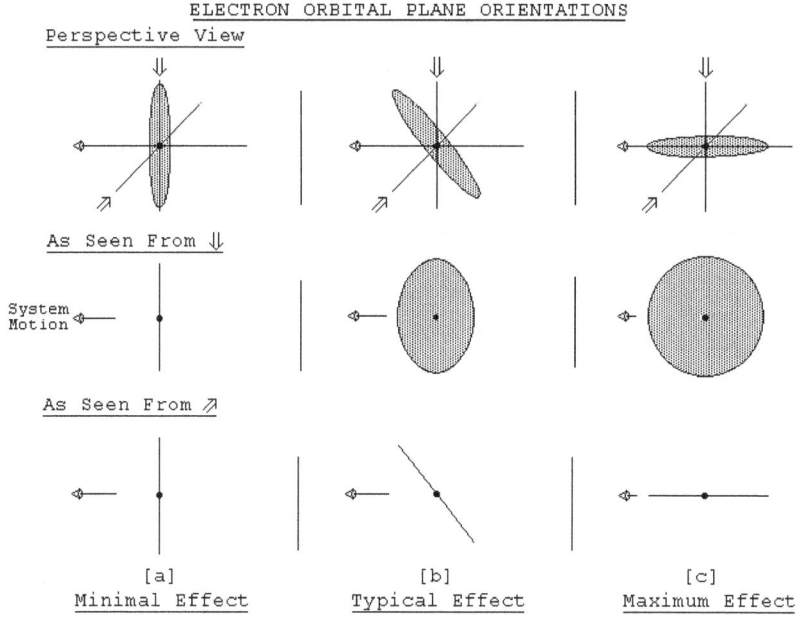

Figure 6-1
Relative Effect of Absolute Motion on Various Orbital Electrons

The figure illustrates different ways that the plane of an orbital electron's orbit can be oriented relative to the absolute motion of the atom's nucleus. If the orbital plane is oriented at right angles to the direction of absolute motion, as in the [a] Minimal Effect column of the figure, then the absolute motion produces the same change in the overall electron resultant speed everywhere in the orbit. The electron's total speed is that resultant. Its orbital speed relative to the nucleus is the circular orbit speed for that orbital shell as already analyzed and presented.

On the other hand, if the orbital plane is oriented parallel to the direction of absolute motion, as in the [c] Maximum Effect column of the figure, then the overall resultant speed of the electron varies between the sum of its circular orbital speed and the absolute motion speed and the difference of the two speeds (see Figure 6-2, below). In general, orbital planes are frequently oriented between those two extremes as illustrated in the [b] Typical Effect column of the figure. For such cases the absolute motion can itself be resolved into two components: one at right angles to the particular orbital plan (Case [a]) and one parallel to it (Case [c]) and the resulting overall effect analyzed in terms of a combination of those two extreme cases.

Figure 6-2, below illustrates the analysis of Case [c] Maximum Effect.

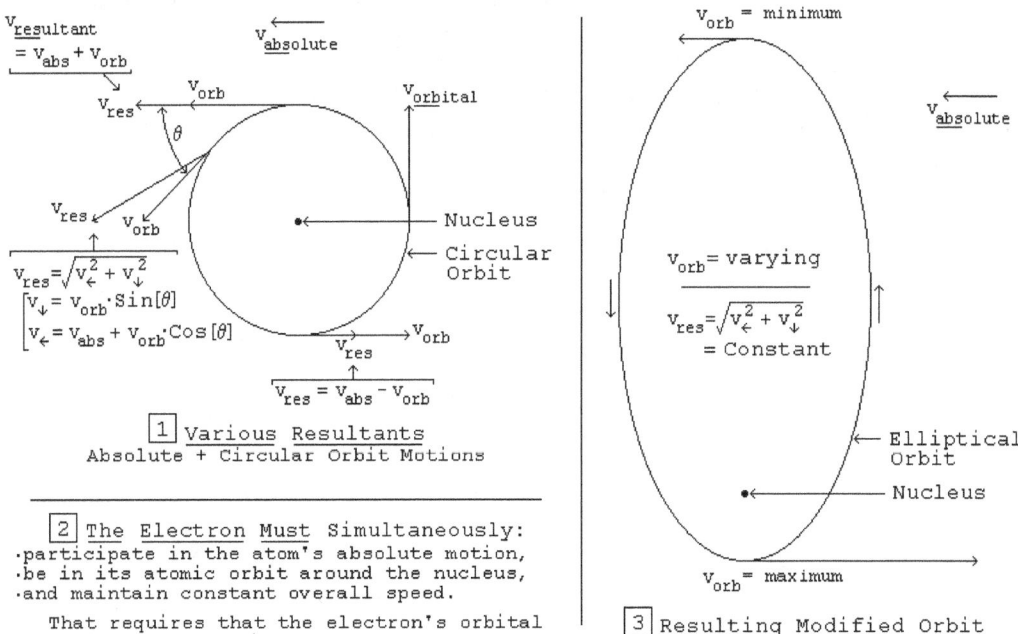

Figure 6-2

The figure is largely self-explanatory. If the electron is in a circular orbit (with consequent constant orbital speed) then the effect of the atom's absolute motion is to vary the electron's absolute speed, which is not acceptable. The only solution, the only *modus operandi*, is for the electron orbital speed to vary so as to compensate for the absolute motion and maintain constant absolute electron speed as shown in box 3 of the figure. The result is elliptical orbits for those orbits in which the orbital plane is not perpendicular to the direction of absolute system motion, that is for those orbits of Cases [b] or [c] or Figure 6-2.

The circular orbit speed in the $n = 1$ orbit of Hydrogen is about $2.2 \cdot 10^6 \cdot m/sec$. Our absolute speed is about $3.70 \cdot 10^5 \; m/sec$. The successive orbit speeds for $n = 2, 3, ...$ are $1/n$ times the $n = 1$ value. Thus the effect of absolute speed and the variations in orbital speeds are quite significant.

It is interesting to recall that the system of orbital quantum numbers developed by 20th Century physics and particularly elaborated by Dirac, used the convention of the projection of an orbital angular momentum vector on a reference axis to define the various orbital tilts. It has now here been found that the tilts are the direct result of the space required for the matter wave of each orbital electron and the tendency of the electrons to space themselves as equidistant from each other as the circumstances permit. And it has now here been found that the "reference axis", an imaginary and missing element in traditional 20th Century physics terms, is actually the orbital plane orientation relative to the atom's absolute motion in space. The $l = 0$ value corresponds to the electron orbital plane being at right angles to the absolute motion, Case [a] of Figure 6-1. The $l = 1$ value produces a Case [b] situation.

That kind of orientation situation is illustrated as an example in Figure 6-3 below. The orbit depicted as horizontal of Figure 6-3 is depicted deemed at right angles to the absolute motion and is therefore circular. The other two orbits of the figure are thus found to be forced to be elliptical, a pair tilted at equal but opposite angles relative to the absolute motion of the atom.

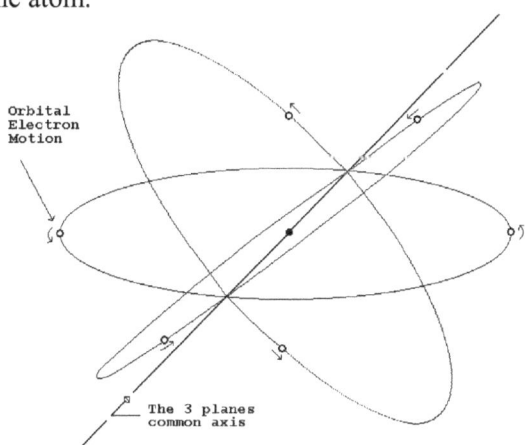

Figure 6-3
Three Orbital Planes and Relative Tilts, n=2 Shell

Returning to the problem of the cause of the fine structure in atomic spectra, there is a second consequence of the orbital electrons' absolute motion. Each electron has a component of magnetic field due to its straight line motion in space in addition to its orbital motion magnetic field. The electron's orbital magnetic field, which is perpendicular to the plane of the orbit, tends to align with the linear motion magnetic field that is due to the atom's absolute motion, which field is circumferential to the electron's direction of absolute motion. There are two possible alignment orientations, that is two orientations when there is no force acting that tends to change the orientation to one of the two. One is orbital motion in the same direction as the absolute motion magnetic field and the other is the opposite. The two differ slightly in energy. It is not "spin-orbit" coupling but "absolute motion - orbit" coupling that operates to produce the fine structure.

The electron's absolute motion magnetic field may seem to be rather weak for the purpose (just as would the magnetic field of a spinning electron so seem), but just as in the hypothesized spin-orbit coupling, both of the actions actually are acting at the same location, that of the electron.

Hyperfine Structure and Spin

High resolution spectral techniques, including the use of tunable lasers, disclose an even more closely spaced splitting of spectral lines which is called hyperfine structure. Analogous to the quantum mechanical explanation of fine structure in terms of hypothesized orbital <u>electron</u> spin, the hyperfine structure is attributed in Quantum Mechanics to <u>nuclear</u> spin, its consequent magnetic field, and its interaction with the electrons. But, the nucleus can no more have a spin magnetic field than can an orbital electron. In this the present physics that is clear for the case of Hydrogen where the nucleus is a proton, a simple *Center-of-Oscillation* however it is also true of all atomic nuclei.

The hyperfine structure stems from electron orbital magnetic field interaction with the magnetic field due to the nucleus' absolute motion in space. Of course, overall the nuclear and orbital electron absolute motion magnetic fields cancel out since the direction of absolute motion is the same but the polarity of the moving charges are opposite. However locally, within the atom there is not general cancellation.

COSMOLOGY AND ASTRONOMY

The Universe's Center and Edge

The physics of relativity has affected cosmology and astronomy by strongly establishing the interpretation or understanding that, in effect, the universe has no edges and no center. The lack of edges is not taken to mean that the universe extends forever; rather it is taken to mean that the space of the universe curves and that if one were to travel long enough in what seems to that traveler to be a straight line he would end up "coming up behind himself where he started".

Furthermore the relativistically conceived universe has no center. In spite of what seems a reasonable assumption, that the center of the universe is where the first moments of the Big Bang took place, that reasonable assumption is denied.

However, the fact of the prime frame of reference, the rest frame, changes all of that.

The prime frame of reference, the rest frame, is the only objective frame. All others are subjective points of view of the observers involved. The fact that our subjective point of view seems reasonable to us, seems "normal" is of no consequence. We live in the prime frame whether we think we can detect it or not; it is the entire universe.

At the instant of the Big Bang a radially outward expansion began and it is still going on. We on Earth are out some large number of light years in one generally radial direction from where that Big Bang started. If we imagine a sphere cut in half to a pair of hemisphere's flat sides facing each other with our outward radius in one of the hemispheres and located where it forms an invisible boundary between two equal quarters making up our hemisphere, then the vast universe we see around us is more or less half of the total. Everything in the other hemisphere is moving away from us and we shell never succeed in having anything to do with it.

So, "our" universe is half or less of all that the Big Bang produced. That doesn't really matter; it is more than enough for us. Yet, just as we are curious to study and learn more about our universe we would like to know about that other half with which we will never interact. The best that we can do is to envision the history of our "half" of the universe.

The mistaken conception that the universe has no edge and no center led to a generally mis-conceived set of conclusions.

- First, is the failure to detect and investigate the universal exponential decay in spite of the substantial evidence for it, see below.

- Second, is the errors in the Hubble Law and theory [see next page]. A number of years ago in about the 1990's the estimates of astronomers and astrophysicists were that the earliest galaxies took about 2½ to 3 billion years to form, that is, that they did not appear until 2.5-3.0 billion years after the Big Bang. Those estimates were based on analysis of the processes involved in star formation and in the aggregation and "clumping" of matter in the early universe.

> Since then improved equipment and techniques [e.g. Keck and Hubble telescopes and gravitational lensing] have resulted in reports of observation of early galaxies having stars that formed as early as 300 million years or less after the Big Bang.

> Such new data has led to the abandonment of the several billion years estimates of the time required for star and galaxy formation and to the adoption of unexplained and unreasonable assertions that early galaxies formed as little as a few hundred million years after the Big Bang. However, an alternative response to those recent data would be to re-examine the Hubble theory from which the age of the universe and the distance to high redshift objects is determined.

- Then, there is the vast error in the age of the universe and in distances to far distant cosmic objects, see Appendix E, "The "Big Bang" Outward Cosmic Expansion".

The Universal Exponential Decay

Since the "Big Bang" the *Propagated Outward Flow* has been gradually depleting the original supply of *medium* in the core of each *Spherical-Center-of-Oscillation*, depleting the amount that drives the *Propagated Outward Flow* through the surface of the core. That process, an original quantity gradually depleted by flow away of some of the remaining quantity, is an exponential decay. Appendix D presents the description of "The Universal Exponential Decay" and the evidence for it.

In the absence of the demonstration that there is a prime frame of reference and the resulting adjustments and conclusions about the universe, the scientific environment was not conducive to investigating the universal decay in spite of the evidence for it. It can be hoped that Appendix D now rectifies that problem.

The Hubble Law

Analysis of the Hubble Law shows that it is asymptotic to an age of the universe that depends on the value of the Hubble Constant. See Appendix E, "The 'Big Bang' Outward Cosmic Expansion", equation *E-41* and Figure E-8. The value of the Hubble Constant is generally taken as in the range of `60 to 75`, but its value remains to be determined. The current generally accepted age of the universe is `13.7 billion years`, which corresponds to a Hubble Constant of `67`. The most recent [2012] determination of a value for the Hubble Constant is `74.3 ± 2.1`.

For high z cosmic objects the Hubble Law results in recession velocities approaching the speed of light. That those velocities are attributed to expansion of space, not to actual velocity of the objects, does not really relieve the problem. According to the Hubble Law the distance between we the observers and those far distant cosmic objects is nevertheless increasing at a rate almost the speed of light which is unreasonable for such immense masses.

The problem of sufficient time after the Big Bang for stars to form, the unreasonable recession velocities implied by the Hubble Law, and even that the Hubble "constant", on which those all depend, is so poorly determined and appears to not be subject to better determination, would all evidence that the Hubble theory is defective and should be replaced.

The "Big Bang" Outward Cosmic Expansion

In the absence of the demonstration that there is a prime frame of reference and the resulting adjustments and conclusions about the universe, the scientific environment was not conducive to investigating the radially outward expansion of the universe from the central source of the Big Bang.

Appendix E rectifies that situation and develops the outward cosmic expansion under the influence of the initial outward explosive energy and the subsequent on-going exponential decay.

The analysis of the prime "at rest" frame of reference produced these useful and beneficial effects of absolute space on science.

More importantly that analysis leads to discovery of the prime universal standard of time and the combination of absolute space and absolute time help resolve a major problem having an adverse effect, not merely on science but on overall human society.

Next: Section 7 – Absolute Objective Time

To address or understand the problem of
Absolute Objective Time
it is necessary to investigate the problem of cause and effect.

Cause and effect are fundamental to the operating of the universe. All of physics is essentially investigation of the role of cause and effect.

And, inherent in the principle of cause and effect is that the cause must precede the effect in the sense that the cause must be extant before the initiation of the effect.

Therefore investigation of cause and effect necessitates investigation of **Time**.

SECTION 7

PRIME OBJECTIVE TIME

Absolute Objective Time

Just as "Invariance" of all physical laws and their constants depends on a single prime frame in which everything is and operates,

so

"Reliability" of cause and effect requires a single prime time in which everything, all events, occur.

THE ABSOLUTE STANDARD OF TIME

The first issue with regard to the subject of absolute objective time is, "Where is it ?" or, more precisely, "What or where is the arbiter, the incontestable standard, the point of reference for objective absolute time."

In sections 2 and 3 it was developed that the fundamental particles, protons and anti-protons, are oscillations at a frequency, f_{wve}, derived from the Big Bang [Equation 3-6]. Specifically the proton and anti-proton oscillations are at the naturally occurring arbitrarily determined frequency of the original oscillation [the wave] that was the beginning of the universe [Equation 2-16].

Even that frequency of oscillation is not always stable in that it varies with variation in the oscillating proton's absolute velocity relative to "at rest", the prime frame of reference as developed in section 4, "Motion and Relativity" [Equation 4-2].

However, all of the universe's protons [the most abundant ubiquitous particle of the universe] when at rest [at zero velocity relative to the prime frame] oscillate at the most fundamental frequency in the universe, f_{wve}, the original oscillation frequency by means of which the material universe began, a frequency that cannot vary except by motion of the proton and which motion is always relative to the fixed, standard prime "at rest" frame of reference.

And that at rest frequency, f_{wve}, the frequency of the proton "at rest", or rather its inverse, the time measured by the period of that oscillation, is a completely stable standard of TIME.

"Time" is the occurring of events in sequence, a sequential order resulting from cause and effect, the dependence of the effect on its cause.

Relative Perceived Time

The Lorentz Contractions, [Section 4 and Equation *4-9*] describe that the time experienced in a frame of reference varies with the velocity of that frame. The Lorentz transformations are coordinate transformations between two inertial frames. More precisely, for frame "A" at velocity "v" as seen by frame "B" relative to frame "B", then observed time periods in frame "A" are lengthened relative to those in frame "B" so that in frame "A" time passes more slowly than time in frame "B".

But, how can that happen ? What is going on ? In Section 4, Equation 4-2, it was shown that the effect of velocity on the *Spherical-Center-of-Oscillation* is to force its frequency of oscillation to decrease and its wavelength to correspondingly increase. That is a result of the essential principle of "Invariance", that the laws of physics and their fundamental constants are the same every where in the universe in every frame of reference. The Equation 4-2 expression for the wavelength increasing is identical in form and constants to the Equation 4-9 expression for the Lorentz Contraction of time. Of course, the "wavelength" of Equation 4-2 is the period of the oscillation in time and that of the Lorentz Contraction.

"How that can happen" and "What is going on" is that the Lorentz Contraction slowing of time is the actual decreasing of every *Spherical-Center-of-Oscillation's* frequency and the corresponding lengthening of its wavelength, the lengthening of its time period, the slowing of the flow of time all forced by its motion, its velocity.

The *Spherical-Center-of-Oscillation's* oscillation produces the stream of pulses of *Propagated Outward Flow* by means of which all Coulomb, gravitational, and magnetic effects take place. The decreasing of its frequency, the time rate at which those pulses occur, is a slowing of the various processes mediated by the Coulomb, gravitational and magnetic action of those pulses, causing those actions to take longer, an apparent slowing of time.

A consequence of the Lorentz variability of time, and of that the rate at which time flows is dependent on the velocity of the frame of reference involved, was the 20th Century denial of absolute objective time. In their lacking the concept of an objective prime frame and of the proton oscillation and its role as the standard of prime objective time the 20th Century scientists' denial of absolute time was wrong but not unreasonable.

In 20th century physics, the relativity of simultaneity is the concept that whether two spatially separated events occur at the same time is not absolute, but depends on the observer's reference frame. That allows variability in time and timing and undermines the fundamental requirement of the physical functioning of the universe, namely that every effect has a precedent cause, that cause and effect always operate, that they are essential and underlie all of physics. The relativity of simultaneity undermines that because it opens to question the fundamental requirement that a cause be extant before the initiation of the effect.

Quantum Mechanics and Time

Quantum Mechanics does not overtly state its denial of cause and effect; however, the interpretations it puts on various effects are equivalent to denial of cause and effect at least for those effects.

One such case is the Quantum Mechanics denial of the limitation of the speed of light in contending the existence of communication or transmission over extensive distances instantaneously. That results from non-understanding of: "what determines the speed of light" and of "the effect of motion on particles".

What Determines The Speed of Light

Every oscillation that we know in nature exhibits, and the very theory of oscillations in the abstract requires, that the oscillation consist of two aspects of the substance which is oscillating [e.g. pendulum position and velocity or electric potential and current] storing and exchanging back and forth the energy of the oscillation. With one aspect varying in oscillatory fashion then when that aspect decreases there must be some "place" for its energy to go, a place in which it is stored until it reappears in that aspect when it increases again. It cannot completely disappear or be lost because the oscillation would die. That "place" is the oscillation's second aspect and it obviously must vary in a manner related to the first aspect's variation with its energy storage in opposite phase.

Like electric inductance and capacitance determining the speed of propagation along a transmission line, μ_0 and ε_0 determine the speed of the *[1 - Cosine]* form oscillation propagation by setting the two aspects of the oscillation in which they are involved, the aspects between which the oscillation energy exchanges back and forth.

But, when the original oscillation came into existence at the beginning of the universe it did so in absolute nothing. There was no "free space" with μ_0 and ε_0. There was nothing but the original oscillation. And, after the immediate explosion into all of the particles of the universe, each of those particles was sending its *Propagated Outward Flow* into nothing, into emptiness.

Where did the *Propagated Outward Flow*'s μ_0 and ε_0 come from? The only thing they could have come from was the original oscillation. There is no other possible source because everything else was absolute nothing, "the zero of existence". The μ_0 and ε_0 are inherent in the substance of the oscillation, which means, μ_0 and ε_0 are also inherent in the outward propagation. Each particle's *Propagated Outward Flow* contains its own μ_0 and ε_0.

And that sets an absolute limit on the speed of communication by *Propagated Outward Flow*, Equation *3-18*.

The Effect of Motion On Particles

With the *Spherical-Center-of-Oscillation* moving in some direction the center's motion and its propagation conflict. In the direction of motion the velocity of the center, *v*, tends to add to the natural value of the speed, *c*, of propagation of the *Propagated Outward Flow* and in the opposite direction it tends to subtract. But, the speed of the flow is fixed; set at *c* by μ_0 and ε_0.

As presented in detail in Section 4, that conflict forces an adjustment of the oscillation of the *Spherical-Center-of-Oscillation*. That adjustment results in increased mass, increased resistance to acceleration, a mass and resistance that approach infinite as the velocity approaches, but never reaches, the speed of light, Equation *4-6* repeated below.

$$(4\text{-}6) \quad m_v = m_r \cdot \frac{1}{\left[1 - \frac{v^2}{c^2}\right]^{\frac{1}{2}}}$$

And that sets an absolute limit on the speed of communication by particle motion.

SUMMARY

At the time of the development of Lorentz's contractions and of Einstein's relativity they had no knowledge of the existence of the prime frame of reference, the "absolutely at rest" frame. They also had no knowledge of the Big Bang and of how it happened and of how it led to the behavior of matter today. Plus, they <u>did</u> have knowledge that perceived or experienced time varied with velocity.

Thus an absolute objective fixed standard of time was literally inconceivable to them. They could only conclude that time was completely relative; it was what each observer independently experienced it to be.

Also, at that time of Lorentz's and Einstein's research the principle of cause and effect was a fully accepted and understood axiom of physics. It was not open to question and was not in need of support.

Their denial of absolute objective time was completely natural and to be expected.

However, now that is all changed:

- There is the now established "arbiter, the incontestable standard, the point of reference for objective absolute time", the oscillation of "at rest" protons, and

- That is the *single prime time in which everything, all events, occur*, and

- Because real objective time of all events is that of the absolute time standard the Reliability of cause and effect is assured.

Time is relative only for various different observers as perceived by them from their personal local frame of reference.

Which means now that it is time to resolve the major problem in human society the problem that has led to world-wide war, rapine and holocausts: the destruction of objective reality and absolute truth.

Section 8 – The Adverse Effect of Subjective Space and Time

SECTION 8

The Adverse Effect of Subjective Space and Time

The 20th Century denial of a prime absolute frame of reference and denial of absolute time left space and time as subjective. That is reality was left subject to the whims and desires of each individual. The results have been catastrophic.

Science on the large scale, that is science dealing with the fundamentals of reality and the universe, has always had and still has a major effect on the non-scientific - social - general philosophic thinking of that science's society and its leaders.

The beginning of the scientific method and the work of scientists such as Copernicus and Galileo resulted in the new period of "The Age of Reason" and "The Enlightenment" – rationality and empiricism replacing dogma and faith.

The new developments that Newton introduced led directly to the concept of the "clockwork universe" and the strong belief in laws, order and regularity.

And, Einstein's denial of objective space and objective time coupled with the 20th Century's attribution of actual uncertainty or indeterminism to all physical objects beyond the original Heisenberg conception, and the advent of Quantum Mechanics with its denial of cause and effect, resulted in our contemporary outlook of a probabilistic reality with no certainty, everything relative with no firm truths.

And, upon that we can lay some of the responsibility for the horrors and tragedies of the 20th Century.

How is that so?

In general, a statement and its contradiction cannot be simultaneously true. Therefore, there are some absolute truths. Thus there is absolute truth, which is the collective body of absolute truths.

Not all statements are absolute truths. Aside from error, which by definition is not true, there is opinion. For example:

- Some people state their liking for candy; some their dislike. It is a matter of opinion.

- But, the statement "Some candy has properties that appeal to some people" is an absolute truth.

The point of view that the questions, "What is truth?" and "What is real?" are meaningless questions without answers is not only incorrect but quite negative and harmful in that it suppresses inquiry and progress that could otherwise take place.

The problem is the general denial of absolute truth and the general acceptance of its contrary – that everything is relative, indeterminate, probabilistic. However, because a statement and its contradiction cannot be simultaneously true there are some absolute truths. Thus there is absolute truth, which is the collective body of absolute truths. This develops in detail as follows.

REALITY

<u>Reality is</u> that which exists, which is. It includes material reality [matter and energy in their various manifestations] and non-material reality [ideas, concepts, feelings, events, *etc.*].

Reality is objective. There can be no subjective reality. The skeptical objections and their refutation are as follows.

- The skeptic who contends that there are different realities for different persons or different situations misunderstands through error in perception or error in judgment. Objective reality is independent of perception and judgment. It exists in itself.

Different persons may <u>experience</u> different personal realities because each experiences a personal sub-set of the comprehensive totality of reality.

- Some skeptics acknowledge the independent objective existence of material reality but contend that ideas and concepts exist only by virtue of minds thinking of them and have no independent objective existence. That contention is in error as follows.

If all minds ceased and subsequently new minds arose, those new minds would develop some of the same ideas and concepts that were in the earlier, now ceased, minds *e.g.*: truth, goodness, right and wrong, beauty, *etc.*, and other abstract concepts such as mathematics and logic. If that ceasing of existing minds and the subsequent arising of new minds were to occur many times over, some of the same fundamental ideas and concepts would reappear in each new set of minds. Such ideas and concepts exist in themselves independently of minds to think of them. They have the same objective existence as does material reality. Some of them are, for example: truth, goodness, justice, right and wrong, love, beauty.

TRUTH

Truth is that which is in agreement with reality. It is objective truth because it corresponds with objective reality. It is absolute truth because there is only one objective reality.

A judgment is a conclusion as to the truth or falsity of a specific statement; that is, a judgment is a conclusion that a specific statement is in agreement with reality [is true] or is not in agreement with reality [is false].

- There can be no subjective truth. Apparent subjective truth results from errors in perception of reality or from errors in judgment as to the agreement with reality, or both. There is only one reality.

- There can be doubt, questions, or issues with regard to specific truths, the doubt arising from insufficient information or from concern as to the validity of the reasoning to reach the judgment. Those problems do not affect objective reality nor objective truth. They only affect our ability to know the specific truths, an effect that can be reduced or removed with better information or better reasoning or both.

- A judgment is conclusively certain if it is impossible for new evidence to change it and its reasoning is beyond criticism. [For example, it is conclusively certain that the sum of the interior angles of a plane triangle is a straight line].

- Otherwise the judgment is in doubt to the degree that those two conditions are not met. The possible states of doubt range from "nearly" or "practically" certain through certain "so far" or "at this time" or "per a preponderance of the evidence" on to the genuinely doubtful. But, such doubt does not change objective reality nor objective truth -- it only describes the limits of our knowledge of the truth.

KNOWLEDGE

Knowledge is accumulated truth. There are two sources or methods to obtaining knowledge: information obtained via the senses [empirical, physical knowledge] and conclusions obtained from logical, rational deduction [metaphysical knowledge]. Both are subject to error; however, that defect is not comprehensive.

The senses may be in error some times through unintended or unrecognized distortions in perception or because of error in our comprehension of that which the senses deliver to us, but the senses are not comprehensively, consistently in error. If our senses were not largely reliable it would be impossible for us to successfully exist. Therefore, while we cannot rely absolutely on the senses [empirical, physical information] as a source of knowledge nevertheless the senses are a valuable and largely reliable source of knowledge.

Likewise, in spite of our best efforts, our logical, rational thinking and analyses can be in error through deficiency in the facts available to us upon which the rationality is based or because of defects in the logic that we apply to the problem. But, our logical, rational thinking is not comprehensively, consistently in error. Again, if our rationality were not largely reliable it would be impossible for us to successfully exist. Therefore, while we cannot rely absolutely on logic and reasoning [metaphysics] as a source of knowledge nevertheless it is a valuable and largely reliable source of knowledge.

How We Can Find Truth

Then, what is the key to accurate, valid, reliable knowledge? The pertinent factors bearing on the validity of any truth, any component of knowledge, are:

- the causality or mechanism involved,

- non-dependence on unsubstantiated assumptions, and

- valid relating to all other truths, to the body of validated knowledge.

These operate as follows.

- <u>Causality or mechanism</u> is apparent from observation and experience which show that every thing and every event has a cause, and that those causes are themselves the results of precedent causes, and *ad infinitum*. Defining and comprehending the causality or mechanism operating to produce any contended or proposed truth is essential to authenticating or validating that truth.

  The candidate truth cannot be deemed valid until its causes and mechanism are analyzed back to an already substantiated operating cause upon which it effectively depends. If that is lacking then it is always possible that a candidate truth will be found not to have a valid precedent operating cause, a valid mechanism in its precedence and, therefore, itself not be valid.

- <u>Assumptions</u> are proposed or contended truths, proposed or contended components of knowledge, that lack sufficient proof or justification to credit them as real truths, as really in agreement with reality. Clearly that infection cannot be part of knowledge without contaminating the whole.

  It is not easy to avoid assumptions. Personal prejudices and beliefs may not be apparent to their holder, or they may be apparent but are nevertheless deemed exceptions to the requirement prohibiting assumptions. That may be because he considers them so important or fundamental as to be beyond question.

  Or it may be because he is psycho-emotionally wedded to them, dependent on them. For example, in the history of philosophy the God assumption appears abundantly, major instances being, for example, Augustine, Aquinas and Descartes.

In the sciences, hypotheses that have not [or not yet] succeeded in advancing to the state of completely determined and validated laws nevertheless acquire over time the status of being treated as if completely validated and not subject to questioning. Major modern instances of this are the "Hubble Constant" and its related cosmology and the irresolvable inconsistency of Quantum Mechanics and Einstein's General Relativity's treatment of gravitation.

In addition there can also be assumptions that are so embedded in the psyche of the pursuer of knowledge that he is not even aware of their presence and influence on his thinking and research.

- <u>Validly relating to the body of validated knowledge</u> is fundamental to what knowledge is: accumulated truth, assembled agreement with reality, that is agreement with that which is. Overall consistency is a fundamental requirement. A component of knowledge not being so compatible would constitute a contradiction, the holding that a thing and its refutation are simultaneously valid.

If those criteria are met then contributions to knowledge produced physically, that is using the senses, or produced metaphysically, that is using reasoning, or produced using both are reliable validated components of knowledge.

Just as there is only one reality and can be only one reality, so is there and can there be only one knowledge, one overall collection of truths, one system of everything.

THE PROBLEM OF ABSOLUTE TRUTH

Truth, being that which is in agreement with reality, is objective truth because it corresponds with objective reality.

It is absolute truth because there is only one objective reality.

The point of view that the questions, "What is truth?" and "What is real?" are meaningless questions without answers is not only incorrect but quite negative and harmful in that it suppresses inquiry and progress that could otherwise take place.

Whether we can know, sense, measure, or understand some aspect of reality or not it still, nevertheless, is.

Its being does not depend on our consent, nor our observation, nor our understanding of it, nor even our own being. We are not gods.

The problem is not whether there is absolute truth or not -- there is. The problem is finding out, coming to know, what the absolute truth is, what is true and what is not. Just what is the "real" reality.

This problem, the difficulty in determining the truth about reality, has beset mankind since the earliest stages of the development of our reasoning. That difficulty -- many have deemed it an impossibility and still do -- has resulted in a more or less collective decision to grant equal validity to a number of different versions of the truth in spite of their being mutually contradictory.

Not that individuals, organizations (e.g. religions, businesses, academia) and governments hold the opinion that their own version of the truth is not correct. Rather, they ardently believe in the correctness of their own views. But, their inability to prove their views and their inability to defeat differing or opposing views necessitates their getting along in some fashion with those other views and the multiplicity of contradictory views of reality.

That state of affairs has existed for so many human lifetimes that it has essentially implanted in our collective and individual thinking the incorrect belief that there is no absolute truth, that truth is what we say it is -- especially that truth is what we can enforce it to be. The contemporary outlook is of a probabilistic reality with no certainty, everything relative, no firm truths.

We have gone from inability to determine the truth to non-belief in its existence and then to belief that truth, and reality, are whatever we believe them to be and can force our fellow (or an organization or government) to accept. The most significant characteristic of the 20th Century, other than its explosion of technology, has been its adoption of the attitude that truth is different for each person and each case, that it is what each individual perceives it to be -- that there is no objective reality, only the subjective reality as perceived by each individual.

The great damage that such thinking does is the license that it gives. It gives license to create, choose, decide upon one's own "reality" and then act accordingly. Such thinking ultimately gives us war, rapine, holocausts.

But, if there is an objective reality, objective truth, then, even if we are not able to completely know and understand it, we are subject to it. We are measured and judged by it; we experience the effects and consequences of it whether we agree and approve or not, and we are compelled to behave accordingly.

Thus objective reality and objective truth,
which indeed exist,
also are desirable and beneficial.

They are, in fact, essential to civilized society.

We can lay some of the responsibility for the horrors and tragedies of the 20th and 21st Centuries upon Einstein's insistence that there is no absolute frame of reference, the probabilistic universe of quantum mechanics, and the attribution of actual uncertainty or indeterminism to all physical objects, an extension far beyond the original valid Heisenberg Uncertainty of measurement due to the act of measurement changing the object measured.

In other words, the problem is the conflict of science with rationality, absolute reality, absolute truth. The resolution of that conflict is as follows.

- x - Einstein's contention that there is no absolute frame of reference offers no causality, no mechanism for its contentions. Thus it lacks one of the fundamental requisites for finding truth.

- √ - It has herein been found [Section 5] that, contrary to Einstein, there is a prime frame of reference, absolute objective space, the "at rest" frame.

x - Quantum Mechanics offers no causality, no mechanism for its contentions of instantaneous communication and denial of causation. Thus it lacks one of the fundamental requisites for finding truth.

√ - It has herein been found [Section 7] that, contrary to Quantum Mechanics, and contrary to Einstein, there is a prime flow of time, absolute objective exact time, the proton oscillation period in the "at rest" frame.

And that absolute time forces every event to be preceded by a cause.

x - The Heisenberg attribution of actual uncertainty or indeterminism to all physical objects offers no causality, no mechanism for its contentions. Thus it lacks one of the fundamental requisites for finding truth.

√ - The finding herein of absolute objective space and absolute objective time, that is absolute objective reality, establishes an exactness to all of the physical universe and to its events and actions which refutes contended or proposed actual material uncertainty and indeterminism.

Thus objective reality,
which is essential to civilized society,
but has been denied for over a century through error,
is now fully restored.

APPENDICES

Appendix A

Why No Immediate Mutual Annihilation

BACKGROUND OF THE PROBLEM

The Big Bang could only have resulted in equal amounts of matter and antimatter for the sake of the principle of conservation as presented in Section 2, *The Origin of Space and Matter* with the assumption that there would have been a complete and almost instantaneous mutual annihilation.

Because that annihilation did not take place it has been hypothesized that the original symmetry was slightly skewed in favor of matter and that the universe is now all matter, all original antimatter having been annihilated with an equal amount of original matter. However that skewed balance conflicts with conservation in the Big Bang.

The Big Bang had to produce equal amounts of matter and antimatter and their total mutual annihilation did not occur because of the conditions there. Rather, while a moderate amount of initial matter / antimatter mutual annihilations may have taken place our present universe contains the remaining matter and antimatter in equal amounts, between some particles of which further mutual annihilations still occur at a modest rate.

The failure of comprehensive matter-antimatter immediate annihilation to occur develops as follows.

CONDITIONS AFFECTING MATTER / ANTIMATTER MUTUAL ANNIHILATION

What Is a Matter / Antimatter Annihilation ?

A positron-electron mutual annihilation, for example, is

(A-1) $\quad _1e^0 + {_{-1}e^0} \Rightarrow \approx + \approx \quad$ where $\approx$ is a photon of gamma radiation

It happens as follows [per Equation A-2].

(A-2) $\quad (_1e^0) + (_{-1}e^0) = U_c \cdot [1 - \cos(2\pi \cdot f_e \cdot t)] - U_c \cdot [1 - \cos(2\pi \cdot f_e \cdot t)]$
$\quad \quad \quad \quad = 0$

The two oscillations literally cancel. The annihilation occurs because the two are point-by-point inverses of each other. Such an annihilation is depicted in Figure A-1 on the following page.

In general for a particular particle and some particular anti-particle of it, their phases and frequencies will not be identical because of their different velocities and histories of relativistic frequency shifts. However, for them to mutually annihilate they must remain co-located for some brief moment sufficient for the event to occur.

For the particles to be co-located for a brief moment their positions and velocities must be identical, which means that their frequencies and their phases will also be identical.

The mutual annihilation energy is the conversion into energy of the entire mass of the two particles involved. The mass of each of the particles is its oscillation [there is nothing else to be the mass]. At annihilation the two particles' oscillations cease to exist by cancelling each other out. Since the center oscillations cease, the last waves of *Propagated Outward Flow* are followed by no flow at all from those centers.

E-M radiation is the propagation of changes in the *Propagated Outward Flow*, changes usually caused by velocity changes of charged particles. The ceasing at annihilation of the oscillations of the two particles involved [the largest change possible] causes a pair of gamma photons, Equation $A-1$, to be propagated.

The photons carry off conservation maintaining energy and momentum. The frequency of each photon is the frequency of the oscillation that just ceased, which corresponds to the mass of the particle. In other words the photon energy, $W = h \cdot f$, is the energy equivalent of the entire mass of the of the particle annihilated.

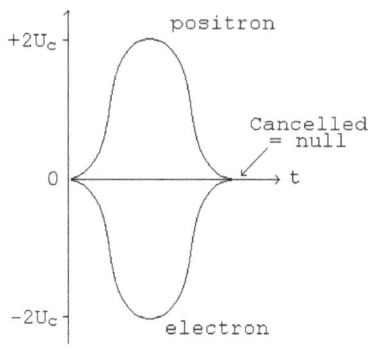

Figure A-1
A Mutual Annihilation

The first issue to investigate is the necessary conditions for a matter / antimatter annihilation to take place: how close must the particle and its antiparticle be and for how long must they remain in such sufficiently intimate contact ?

In addition to those two factors there is the more obvious requirement that the two particles involved be true antiparticles of each other [for example, a proton and an antiproton or an electron and a positron, but not a proton and a positron nor a proton and an electron]. Furthermore in general, particle / antiparticle annihilations are relatively unlikely between electrically neutral particles [for example, a neutron and an antineutron] because the only effects tending to bring the two together are their very weak gravitational attraction or chance encounter.

The Closeness Criterion

Indication of how close the two participating particles must be for their annihilation to take place can be found from the decay of a free neutron [not one that is part of an atomic nucleus] into a proton and an electron, a natural process with a mean lifetime before decay of about 881.5 seconds. For the neutron decay to be successful the proton and electron product particles must derive from the parent neutron not only their rest masses but also sufficient kinetic energy so that they are at escape velocity relative to each other, else they would be attracted back together and recombine. [One can neglect the also emitted electron anti-neutrino which is of zero or negligible mass.]

The escape velocity of the two particles is, at first consideration, an awkward problem because the separation distance of the two particles, which appears in the denominator of the expression for their Coulomb attraction, would seem to be required to be as small as zero. That is, at first consideration the escape velocity required is infinite. But, since infinite escape velocity is impossible yet the escape occurs, then the starting point, the minimum separation distance that can occur must be greater than zero. In other words, the neutron decay products, a proton and an electron, exist as such only when separated by some minimum Separation Distance, s, and their state at lesser separation distances appears as their parent neutron.

Therefore, since if the proton and the electron are separated by less than that minimum distance they do not exist as proton and electron but rather as the neutron, and at separation distances greater than that minimum they are the pair of separate particles, then that Separation Distance is a measure of how close a proton and an electron must be to unite into a neutron and is indicative of the spacing at which a particle and its antiparticle mutually annihilate.

The point is that the excess of the mass of the neutron over that of a proton plus that of an electron must supply the proton and electron relativistic kinetic masses needed to escape the decaying neutron. The detailed analysis and relativistic calculations can be found in Appendix A-1, *The Neutron*. The results are as follows.

(A-3) - The escape velocities:

$$v_e = 275,370,263. \text{ meters per second}$$
$$= 0.918,536,33 \cdot c$$
$$v_p = 379,350.6975 \text{ meters per second}$$
$$= 0.001,265,378 \cdot c$$

- The minimum Separation Distance:

$$S = 1.3 \cdot 10^{-15} \text{ meters}$$

Some years ago experiments involving measurement of the scattering of charged particles by atomic nuclei, yielded an empirical formula for the approximate value of the radius of an atomic nucleus to be

(A-4)   Radius = $[1.2 \cdot 10^{-15}] \cdot$ [Atomic Mass Number] meters

which formula would indicate that the radius of the proton as a Hydrogen nucleus (atomic mass number $A = 1$) is about $1.2 \cdot 10^{-15}$ meters.

The mass of the proton can be expressed as an equivalent energy, $W_p = m_p \cdot c^2$, and that as an equivalent frequency, $f_p = m_p \cdot c^2/h$, or as an equivalent wavelength, $\lambda_p = c/f = h/m_p \cdot c$. That wavelength (not a "matter wavelength") for the proton is

(A-5) $\lambda_p = 1.321,410,0 \cdot 10^{-15}$ meters

quite near to the empirical value for the proton radius from Equation A-4 and the Separation Distance, s, of Equation A-3. Thus the Separation Distance boundary between a proton and an electron as separate particles versus combined into a neutron is about 1 proton radius, the equivalent wavelength for the proton mass per Equation A-3.

Then for a proton and an antiproton the boundary between their being the two separate particles and their mutually annihilating is a proton radius, a Separation Distance of $s_p = \lambda_p = 1.321,410,0 \cdot 10^{-15}$ meters. At that boundary if their velocities have a sufficient net component directly toward each other [per the time criterion, below] they would seem to be able, and likely, to mutually annihilate, and otherwise the annihilation would seem not possible.

Similarly, the mass of the electron or the positron can be expressed as the equivalent energy, $W_e = m_e \cdot c^2$, and that as its equivalent frequency, $f_e = m_e \cdot c^2 / h$, or equivalent wavelength, $\lambda_e = c/f = h / m_e \cdot c$. That wavelength (not a "matter wavelength") for the electron / positron is

(A-6) $\lambda_e = 2.426,310,6 \cdot 10^{-12}$ meters.

Then for an electron and a positron the boundary between their being the two separate particles and their mutually annihilating is a Separation Distance of $s_e = \lambda_e = 2.426,310,6 \cdot 10^{-12}$ meters. At that boundary if their velocities have a sufficient net component directly toward each other [per the time criterion, below] they would seem to be able, and likely, to mutually annihilate, and otherwise the annihilation would seem not possible.

Then, what is that sufficient net velocity?

The Time Criterion

The mutual annihilation of a particle and its antiparticle is symbolized as in the following example for a proton and an antiproton.

(A-7) $_1p^1 + {}_{-1}p^1 \Rightarrow \gamma + \gamma$ where γ is a gamma photon

In the present case of a proton and an antiproton the mass of each of the protons is converted into the energy of the related γ photon. The frequency and period of each of those two photons is as follows.

(A-8) $f_{\gamma p} = m_p \cdot c^2 / h$

$T_{\gamma p} = 1/f_{\gamma p} = h/[m_p \cdot c^2] = 4.407,749,3 \cdot 10^{-24}$ seconds

In communications theory it is shown that a sinusoidal oscillatory signal must be sampled at least twice per cycle for the signal to be correctly represented. That is, two independent datums are required so as to determine the value of the oscillation's two absolute parameters, its amplitude and its frequency. [It's phase is relative, not absolute.] That implies that the time duration of a proton / antiproton mutual annihilation must be the period of each of the resulting photons.

(A-9) $\Delta t_{proton \,/\, antiproton} = T_{\gamma p} = 4.407,749,3 \cdot 10^{-24}$ seconds

Similarly for an electron / positron mutual annihilation, the time duration would be

(A-10) $\Delta t_{electron/positron} = T_{\gamma e} = 8.093,301,0 \cdot 10^{-21}$ seconds.

While those are very brief times they are not instantaneous.

In the case of a particle and its antiparticle coming together from significantly far apart, the particles will have accumulated significant velocity toward each other by the time they arrive at Separation Distance s because of having been accelerated by their mutual Coulomb attraction. However, the situation was different for the Big Bang.

Why the Criteria Failed in the Case of the Big Bang

The number of particles resulting from the original Big Bang is estimated to have been about 10^{85} [Appendix B, *The Limitation of the Original Envelopes*], and those particles emerged on paths that were initially radially outward. The event was overall spherically symmetrical on the large scale, but at the local particle level perfect symmetry was impossible because of the nature of finite particles versus a smooth non-particulate substance. Initially all of the particles were on divergent paths although for two adjacent particles the amount of the divergence was minute.

For a proton and an adjacent antiproton in the Big Bang to be separate [not annihilated] at the instant of being projected outward in the Big Bang, they had to be separated by at least the above-developed $S_p = 1.321,410,0 \cdot 10^{-15}$ meters. For them to then annihilate their Coulomb attraction would have had to accelerate them into co-locating in the required time criterion starting from their initially zero velocity toward each other. [Actually they would have had non-zero but minute velocities away from each other because each follows its own outward radial path.] The issue is whether their Coulomb attraction can accelerate the two particles to the point of co-locating within the time frame of Equation A-9 [or Equation A-10 for an electron / positron case].

If, for example, for their mutual annihilation, the proton or the antiproton is to travel <u>at constant velocity</u> its half of the separation distance, ½·S_p, in time $T_{\gamma p}$, so as to be co located with its antiparticle at the end of that time, it would require a speed of

(A-11) $$v_p = \frac{\text{½} \cdot S_p}{T_{\gamma p}} = 0.5 \cdot c \quad \text{[half light speed]}$$

and if the electron or the positron, for their mutual annihilation, is to travel its half of the separation distance, S_e, in time ½·$T_{\gamma e}$ <u>at constant velocity</u> it would require a speed of

(A-12) $$v_e = \frac{\text{½} \cdot S_e}{T_{\gamma e}} = 0.5 \cdot c \quad \text{[half light speed]}.$$

The achieving of that speed, if even only by the very end of the extremely short time period of the acceleration and travel, 10^{-21} seconds or less, would be difficult. The particles moving continuously at that <u>constant velocity</u> throughout their travel from separated to co-located is impossible in that they commence their travel of distance s from essentially zero velocity toward each other.

Furthermore, the analysis of the Coulomb interaction at close separation distances presented in Appendix A-1, *The Neutron* shows that the attraction weakens drastically at close quarters per Figure A-2, below, reproduced from that Appendix A-1.

[The figure shows the form of the reduction in the Coulomb attraction as a function of the charge separation radial distance relative to a proton mass equivalent wavelength, λ_p.]

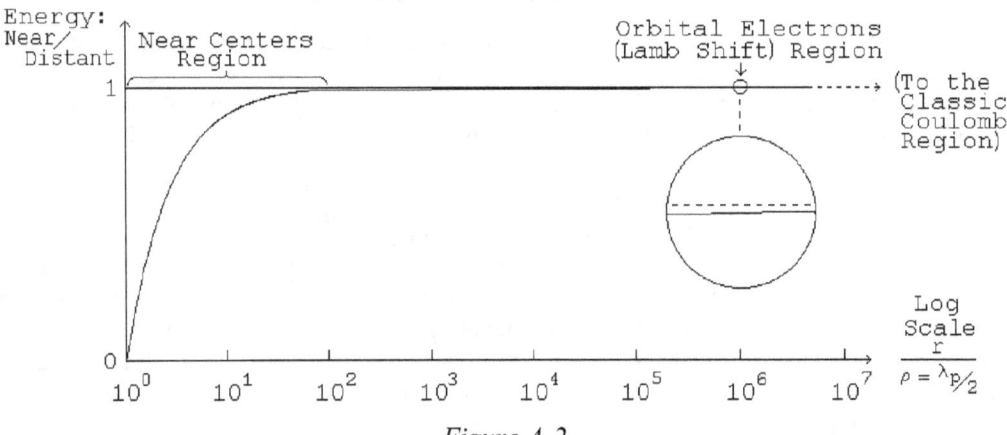

Figure A-2
Coulomb Effect <u>Reduction Factor</u> When Charges Are Near to Each Other

Finally, the posited particle and its antiparticle, emerging from the Big Bang, with spacing adjacent to each other as closely as possible, and on radially outward paths, were not alone. They were surrounded by a more or less uniform, symmetrical, large group of like particles and antiparticles. Any Coulomb tendency to unite the posited particle pair was largely offset by the similar tendency of each member to unite with the adjacent particle on its other side. The net Coulomb action on a specific particle or antiparticle was certainly insufficient to produce enough acceleration to enable the particle to transit its half of the Separation Distance in the required gamma photon period.

In summary:

- Adjacent Big Bang product particles and their antiparticles,

- Initially spaced optimally for co-locating [as closely as possible yet independently separate],

- Traveling outward at near light speed on essentially parallel paths [actually minutely diverging paths],

- Are unable to accelerate toward each other, from zero initial such velocity, quickly enough for their annihilation to produce the known actual gamma photons that would have to result from their mutual annihilation.

- That is, they cannot travel to the point of annihilation in time for the annihilation gamma photons to be the correct frequency to carry off the energy equivalent of the input particles, the pre-annihilation proton / antiproton or electron / positron.

In other words a Big Bang mutual annihilation was much more difficult, and rare, than one might have assumed. A large scale annihilation of matter and antimatter could not have taken place in the Big Bang. The result is that the present universe contains both matter and antimatter in equal amounts because of the original symmetry.

A Universe Containing Both Matter Regions and AntiMatter Regions

Why Matter and AntiMatter Regions Are Able to Co-Exist

Of course, matter / antimatter mutual annihilations in general are not as awkward as they were for the original Big Bang with its peculiar initial conditions. Of interest here, however, is the case of the interstellar medium. It is the interstellar medium that must be examined because it is the natural boundary between regions of matter and regions of antimatter; where, if they are to occur, the anticipated matter / antimatter annihilations should be occurring and yielding their looked-for gamma ray flux.

In the interstellar [and intergalactic] medium the particles and antiparticles start from being significantly separated, residing in the vacuum of interstellar space, which vacuum, while not devoid of competing particles, has a much lower particle density than the original Big Bang. They do not suffer the disadvantage of being in a dense milieu of particles and antiparticles whose Coulomb attractions tend to cancel out their effects. And, they avoid the disadvantage of always starting their mutual Coulomb attraction toward each other with no initial velocity. Without regard for any mutual attraction between particular particles and antiparticles, they all move with significant velocities.

However, those velocities are in general not oriented toward the combination of a pair. Rather, the velocity directions are a combination of [a] some component distributed randomly over the particles in essentially all possible directions, and [b] some amount corresponding to a general flow direction.

Table A-1, below summarizes the particle [and antiparticle where applicable] content of interstellar space. The density of the particles, and their related mean distance apart are such as to militate against any significant number of encounters, whether aided by Coulomb attraction or not. [Excepting solar wind, which is local to star's nearby environment, most of the interstellar medium is Hydrogen atoms, not ions.] [Gravitation can be ignored here, it being decades of orders of magnitude weaker than Coulomb attraction.]

Region	Size	Particle Density [/cc]	Particle Energy
Our Solar Wind	Sun Neighborhood	10.	0.001- 0.004 × c
Our Local Cloud	60 Light Years	0.1	~ 7,000 °K
Our Local Bubble	300 Light Years	0.001	~ 1,000,000 °K
Intergalactic Space	[The Universe]	0.000 ... ?	?

Table A-1 – The Interstellar Medium

As has been pointed out in analyses of our solar wind, with typically *1 atom* in each *10 cm³* of interstellar gas in our local cloud and *10 ions* in each *cm³* of our solar wind, the particles are so far apart that the solar wind and interstellar gas flow through each other without being disturbed by collisions. On that basis, the even less dense regions of the interstellar medium such as ones like our local bubble, those within galaxies in general, and those in intergalactic space are even less conducive to particle / antiparticle encounters.

Another factor bearing on the likelihood of matter / antimatter mutual annihilations occurring in interstellar space is as follows. Because gravitational and Coulomb field attraction communicate at c, particles are attracted to where the attractor was, not where it is. That tends to produce orbital motion or "sling shot" non-collision passages rather than direct collisions. For example, a proton traveling at $0.000,001 \cdot c$ [only 300 meters/second] and at a distance of 0.001 millimeter from another charged particle [compare that distance with the spacing implied by the densities of the above table] will travel a distance equal to 757 of its proton radii during the time that its Coulomb field communicates at velocity c to the other charged particle its then Coulomb attraction impulse.

All of these various factors taken into account, matter / antimatter collisions must be quite infrequent events in the interstellar medium. When such mutual annihilations occur the appropriate gamma photons are emitted.

Indications of Some Matter / AntiMatter Mutual Annihilations

A most likely indication of our detection of cosmic matter / antimatter annihilations is Gamma Ray Bursts [GRB's].

GRB's are flashes of gamma rays coming from seemingly random places in deep space at random times. GRB's last from milliseconds to minutes, and are often followed by "afterglow" emission at longer wavelengths. Gamma-ray bursts are detected by orbiting [*Swift*] satellites about two to three times a week. All known GRB's come from outside our own galaxy. Most GRB's come from billions of light years away [as much as $z = 6.3$ or more].

Under the assumption that a given burst emits energy uniformly in all directions, some of the brightest bursts correspond to a total energy release of 10^{47} joules, nearly a solar mass converted into gamma-radiation in a small amount of time. No candidate process other than a significant matter-antimatter annihilation is able to liberate that much energy so quickly.

Appendix A-1

The Neutron

The fundamental, basic and most simple particles, the proton and the electron and their anti-particles, are developed in the preceding Section 3, *The Behavior of Matter*. The other usually stable particles, the atomic nuclei and the neutron, are combinations of those basic particles.

The evidence that the neutron is a combination of an electron and a proton is overwhelming.

- Unlike the case with atomic nuclei, where the presence of multiple protons and their mutual electrostatic repulsion makes the nucleus tend to fly apart, an electron and a proton would tend to bind together in mutual electrostatic attraction. No binding energy or mass deficiency would be needed for an electron - proton combination.

- This correlates with the neutron mass, which exceeds the sum of the masses of the hypothesized components, a proton and an electron, by `0.000,839,854 amu` (more than the mass of an electron). The neutron has in this sense a negative mass deficiency or binding energy, a mass excess. One would expect this since the act of combining a proton and an electron should also include at least some of the energy of their mutual attraction.

- Because of the negative binding energy one would expect the neutron to be unstable. While it is stable in a stable atomic nucleus, where it is affected by its overall nuclear environment, free on its own it readily decays into a proton and an electron with a mean lifetime before decay of about `881.5` seconds.

- Of course the combination naturally yields the neutron's electrostatic neutrality.

The primary traditional objection to the concept stems from the matter wave wavelength of the electron. In that view the wavelength associated with the electron component of the proton-electron combination would be far too large and in direct contradiction to observed cross-sections and wavelengths.

However, that objection applies to a "bunch of grapes" concept of the two particles' combination – their, so to speak, sitting side by side like two peas in a pod. But if the two particles combine more intimately into a new neutron form their waves combine more intimately. Figure A-1-1, below, shows the combination of two oscillations at very different frequencies, the higher representing the proton and the lower, the electron of the proton - electron pair of which a neutron would be composed.

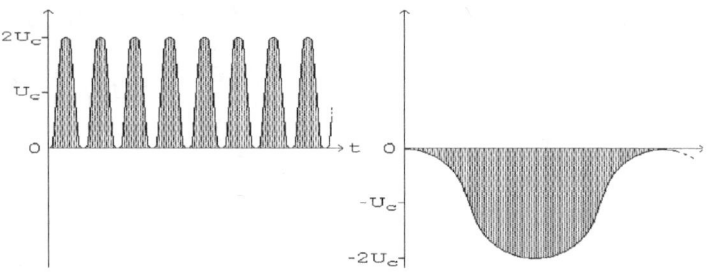

Figure A-1-1(a)
Proton & Electron Oscillations, Two Different Frequencies

As in Figure A-1-1(b), below while our eyes can perceive the longer wavelength in the combined wave form (the envelope), the actual oscillation is only at a wavelength essentially that of the shorter input wavelength. The electron's matter wave need not be a problem.

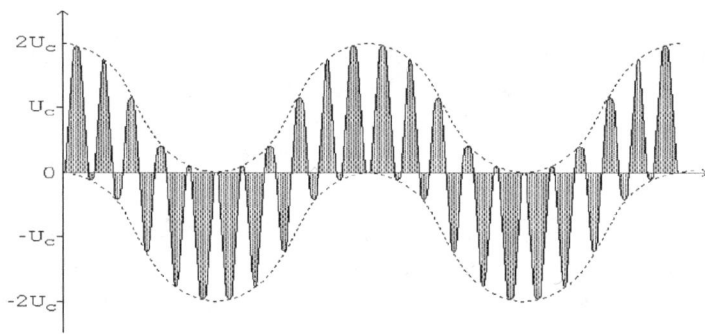

Figure A-1-1(b)
The Sum Oscillation, The Neutron

The *Spherical-Center-of-Oscillation* equation of the neutron, depicted in Figure A-1-1 above, is the sum of those equations of the electron and the proton.

$$(A\text{-}1\text{-}1) \quad U(_0n^1) = U_c \cdot \left[1 - \cos(2\pi f_p t)\right] - U_c \cdot \left[1 - \cos(2\pi f_e t)\right]$$
$$= U_c \cdot \left[\cos(2\pi f_e t) - \cos(2\pi f_p t)\right]$$

The masses of the proton and electron the combination of which is the neutron are not their rest masses even though their combination in the neutron yields the neutron's rest mass. The component masses are the particles' relativistic masses at high velocity. This comes about as follows.

Since a neutron naturally decays into a proton and an electron those decay particles must be emitted at a velocity equal to or greater than their escape velocity. That is, because the proton and electron strongly mutually attract each other electrically, unless they separate at their mutual escape velocities they will immediately re-combine into a neutron.

Put another way, for a neutron to be formed from a proton and an electron the two must come together from the state of being mutually independent of each other. That means that they must mutually accelerate toward each other. In so doing they will each be at escape velocity and have the kinetic energy of that escape velocity at the moment of their combining into the new particle, the neutron.

A-1 – THE NEUTRON

The portion of the neutron's overall rest mass that corresponds to the component proton and electron's escape velocity kinetic energy is the neutron rest mass less the sum of the proton and the electron rest masses.

$$(A\text{-}1\text{-}2) \quad \Delta m_n = m_{neutron,\ rest} - [m_{proton,\ rest} + m_{electron,\ rest}]$$

$$= 1.008{,}664{,}904 - \ldots$$
$$\ldots - [1.007{,}276{,}470 + 0.000{,}548{,}579{,}903]$$
$$= 0.000{,}839{,}854 \text{ amu}.$$

In the "classical" sense escape velocity refers to an object of some mass that is gravitationally bound to some other mass, for example a rocket to be launched from Earth. The force attracting the two objects, the rocket and the Earth, to each other acts on them equally in magnitude and opposite in direction. Consequently, momentums that are equal in magnitude and opposite in direction are imparted to them. Since momentum is the product of mass and velocity, when one object (Earth) is much more massive than the other (the rocket) it may be assumed with negligible error that it (the Earth) is not accelerated and its velocity is negligible. Then all of the kinetic energy is attributable solely to the rocket. That kinetic energy must be equal to the gravitational potential energy binding the rocket to the Earth (the two to each other) for the rocket to escape the Earth's gravitational pull.

However, in the case of a proton and an electron the assumption that only the particle of lesser mass is accelerated and that the other particle's kinetic energy is negligible is not valid. It is not that the electron escapes from the proton; they escape from each other. Or, it is not that the electron falls toward the proton; they fall toward each other. The kinetic energy of each is involved and the sum of the kinetic energies must equal or exceed the binding potential energy for the velocities to be at or in excess of escape velocity.

The analysis is as follows (where r is the closest separation between the escaping objects or particles).

(A-1-3)

Gravitational	Electrostatic
Rocket [R] escapes from from Earth [E]	Proton [p] and electron [e] escape each other

(a) $PE = Force \cdot r$

$PE = \left[G \cdot \dfrac{m_R \cdot m_E}{r^2} \right] \cdot r$	$PE = \left[\dfrac{q_p \cdot q_e}{4 \cdot \pi \cdot \varepsilon_0 \cdot r^2} \right] \cdot r$

(b) Final (escape) Kinetic Energy (KE) = Initial Potential energy (PE)

$KE_R = PE_{total}$	$KE_p + KE_e = PE_{total}$
$\tfrac{1}{2} \cdot m_R \cdot v_R^2 = G \cdot \dfrac{m_R \cdot m_E}{r}$	No direct solution
	A 2nd relationship is $\|P_p\| = \|-P_e\|$ P is momentum
$v_{R,esc} = \left[\dfrac{2 \cdot G \cdot m_E}{r} \right]^{\frac{1}{2}}$	The two relationships must be simultaneously solved for the velocities

For the gravitational case the escape velocity formulation does not involve the mass of the escaping object. In that sense it is independent of the relativistic mass increase with velocity. Furthermore, in the usual cases treating escape velocity of objects (rocketry and astronautics) the velocity never approaches magnitudes at which significant relativistic effects occur.

However, for the electrostatic case the escape velocity formulation must include the masses of the particles, which masses themselves can vary with their velocity. And, in the case of particles, velocities large enough to involve relativistic effects are likely to occur. Therefore, the electrostatic case must be treated relativistically. The simultaneous solution of the electrostatic case's two equations, kinetic energy and momentum, is as follows.

(A-1-4) <u>Momentum</u>

Magnitude of Proton Relativistic Momentum = Magnitude of Electron Relativistic Momentum

$$\frac{m_p}{\left[1-\frac{v_p^2}{c^2}\right]^{\frac{1}{2}}} \cdot v_p = \frac{m_e}{\left[1-\frac{v_e^2}{c^2}\right]^{\frac{1}{2}}} \cdot v_e \qquad m_p \text{ \& } m_e \text{ are rest masses}$$

Solving the above for v_p the following is obtained.

(A-1-5) [<u>Momentum</u> continued]

$$v_p = \frac{m_e \cdot v_e}{m_p \cdot \left[1-\frac{v_e^2}{c^2}\right]^{\frac{1}{2}}} \cdot \frac{1}{\left[1+\frac{m_e^2 v_e^2}{c^2 \cdot m_p^2 \cdot \left[1-\frac{v_e^2}{c^2}\right]}\right]^{\frac{1}{2}}}$$

(A-1-6) <u>Energy</u>

Relativistic Energy [As Mass] Is Conserved

$$\left[\frac{KE_p + KE_e}{c^2}\right]_{gained} = \left[\frac{PE_{total}}{c^2}\right]_{lost}$$

$$\left[m_{p,v} - m_{p,rest}\right] + \left[m_{e,v} - m_{e,rest}\right] = m_n - \left[m_{p,rest} + m_{e,rest}\right] \equiv m_{n,\Delta}$$

$$\left[\frac{m_p}{\left[1-\frac{v_p^2}{c^2}\right]^{\frac{1}{2}}} - m_p\right] + \left[\frac{m_e}{\left[1-\frac{v_e^2}{c^2}\right]^{\frac{1}{2}}} - m_e\right] = m_{n,\Delta}$$

The above equations treat the excess of the neutron's rest mass above the sum of the rest mass of a proton plus that of an electron to be the relativistic KE gained by the two particles in approaching each other from infinite separation distance [per the concept of "escape velocity"].

A-1 – THE NEUTRON

The issue here is: how far apart are the proton and electron in their collision paths toward each other when they have the above kinetic masses, $m_{p,v}$ and $m_{e,v}$? For the calculations to be correct, that is for the hypothesis to be correct, their separation distance at that moment must be such that the two colliding particles are exactly at the moment of combining into the neutron. First the velocities, v_p and v_e, will be calculated by the simultaneous solution of Equations A-1-4 and A-1-5. Then the separation distance of the two particles at the moment of collision will be determined.

(A-1-7) Simultaneous Solution of Equation A-1-4 With A-1-5

The expression for v_p from Equation A-1-4 is substituted for v_p in the denominator of the first term of the expression obtained in Equation A-1-5. The resulting expression has only v_e unknown and is solved for that value.

Rather than manipulating that expression a computer aided design program is used to calculate selected trial values of v_e until the correct result for $m_{n,\Delta}$ [$m_{n,\Delta} = m_n - m_{p,rest} - m_{e,rest}$] is obtained.

The results of that process are as follows.

(A-1-8) v_e = 275,370,263. m/s

$\qquad$ = 0.918,536,33 · c

$\quad v_p$ = 379,350.6975 m/s

$\qquad$ = 0.001,265,378 · c

At those velocities the proton and the electron have total (relativistic) masses of

(A-1-9)
$$m_{e,total} = \frac{m_{e,rest}}{\left[1 - \frac{v_e^2}{c^2}\right]^{\frac{1}{2}}} = 2.529{,}490{,}15 \cdot m_{e,rest}$$

$$= 0.001{,}388{,}308{,}25 \text{ amu}$$

(A-1-10)
$$m_{p,total} = \frac{m_{p,rest}}{\left[1 - \frac{v_p^2}{c^2}\right]^{\frac{1}{2}}} = 1.000{,}000{,}80 \cdot m_{p,rest}$$

$$= 1.007{,}276{,}596 \text{ amu}$$

and their sum is the mass of the neutron.

(A-1-11) $m_{neutron} = m_{p,total} + m_{e,total}$

$\qquad$ = 1.007,276,596 + 0.001,388,308,25

$\qquad$ = 1.008,664,904 amu

(These calculations assume that the component proton and electron are in a state of zero momentum and zero kinetic energy before being mutually accelerated into each other. It likewise assumes that the resulting neutron has zero kinetic energy and zero momentum because all the components' kinetic energy goes entirely into the neutron's rest mass and the two component's momentums are equal and opposite in direction netting to zero in combination. To the extent that the components do have initial kinetic energy and momentum then either the resulting neutron will have kinetic energy equal to the sum of the components' initial kinetic energies and momentum equal to the net of the two components' initial momenta or some of those quantities may appear in the form of some type of neutrino given off at the time the particles combine.

(Likewise, in describing the decay of a neutron into a proton and an electron, it was assumed that the neutron initially had zero kinetic energy and zero momentum. To the extent that that is not the case then some form of neutrino will account for the kinetic energy and net momentum not accounted for by the decay product proton and electron.)

THE REMAINING ISSUE IS:
HOW FAR APART ARE THE PROTON AND ELECTRON IN THEIR COLLISION PATHS TOWARD EACH OTHER WHEN THEY HAVE THE ABOVE KINETIC MASSES ?

Their separation distance at that moment must be such that the two colliding particles are exactly at the moment of combining into the neutron.

An initial calculation of that separation distance, r, is as follows.

(A-1-12)
$$[\text{Potential Energy}_{\text{As Mass}}] \equiv \frac{PE}{c^2} \text{ and must} = m_{n,\Delta}$$

$$\frac{PE}{c^2} = \frac{q_{proton} \cdot q_{electron}}{4\pi \cdot \varepsilon_0 \cdot r} \cdot \frac{1}{c^2} = [0.000,839,854 \text{ amu}] \cdot [^{kg}/_{amu}]$$

$$r = \frac{q_{proton} \cdot q_{electron}}{4\pi \cdot \varepsilon_0 \cdot c^2} = \frac{1}{[0.000,839,854 \text{ amu}] \cdot [^{kg}/_{amu}]}$$

$$r = 1.840,636,27 \cdot 10^{-15} \text{ meters.}$$

Some years ago experiments involving measurement of the scattering of charged particles by atomic nuclei, yielded an empirical formula for the approximate value of the radius of an atomic nucleus to be

(A-1-13) Radius = $[1.2 \cdot 10^{-15}] \cdot$ [Atomic Mass Number] meters

which formula would indicate that the proton radius (atomic mass number $A = 1$) is about $1.2 \cdot 10^{-15}$ meters.

The mass of the proton can be expressed as an equivalent energy, $m \cdot c^2$, and that as an equivalent frequency, $m \cdot c^2 / h$, or an equivalent wavelength, $h / m \cdot c$. That wavelength (not a "matter wavelength") for the proton is

(A-1-14) $\lambda_p = 1.321,408,96 \cdot 10^{-15}$ meters

quite near to the empirical value for the proton radius from Equation A-1-11.

Thus the initial calculation of the separation distance of the proton and electron when their kinetic masses are just correct for them to form a neutron, Equations A-1-9, A-1-10 and A-1-11, results in a separation distance of about 1½ proton radii or equivalent wavelengths, Equation A-1-12. That uncorrected result is so close as to essentially validate that the neutron is a combination of a proton and an electron.

However, there is more.

The result at Equation A-1-12 must be corrected for a variation in the magnitude of the classical Coulomb interaction as the charges approach near to each other. The direction of the electrostatic effect of a charge is radial to the charge location. At great distances from a charge all of those radii in a local sample are such a small part of the total spherical Coulomb action that they are effectively parallel. But, near to the charge they all effectively diverge (as, of course, they actually do in all cases). That reduces the electrostatic force and requires the charges to approach each other more closely than to the distance calculated at Equation A-1-12 - in fact to a separation distance of λ_p exactly, within the limitations of the precision of our data. This develops as follows.

When the two charges are relatively near to each other there is less Coulomb effect because of the radial direction of the Coulomb effect action relative to the charges. Coulomb's law, expressed as potential energy as in Equation A-1-12, above, now becomes as follows.

(A-1-15)
$$[\text{Potential Energy}_{\text{As Mass}}] = \frac{[\text{Reduction Factor}] \cdot PE}{c^2}$$

$$= [\text{Reduction Factor}] \cdot \frac{q_{proton} \cdot q_{electron}}{4\pi \cdot \varepsilon_0 \cdot r} \cdot \frac{1}{c^2}$$

and must $= m_{n,\Delta} = [0.000{,}839{,}854 \text{ amu}] \cdot [\text{kg/amu}]$

But, what is the formulation for the *Reduction Factor*?

For the analysis of the effect of the two charges being so near to each other that the radial divergence of the rays is significant the illustration and dimensions of Figure A-1-2, below, are used. In order to be useful the figure is greatly exaggerated, that is α, β, d and so forth are actually too minute to be seen in an unexaggerated figure.

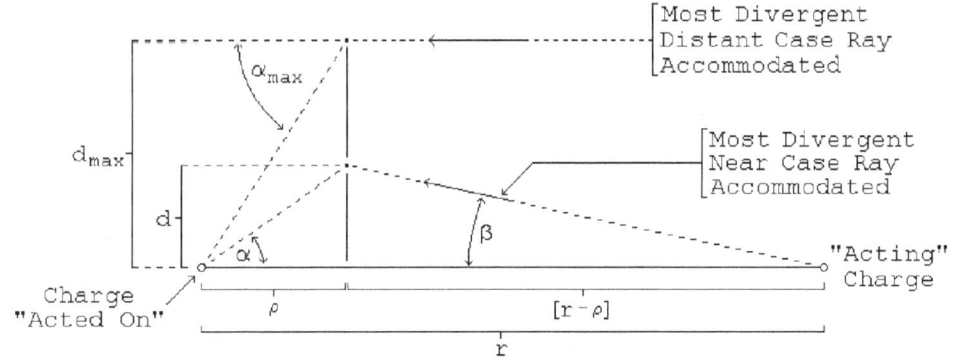

Figure A-1-2
Analysis of Case of Charges Close to Each Other

Even in the case of charges that are far apart, there is only one single ray that is a straight line from one charge to the other. All other rays of electric field must diverge at least minutely from that one straight ray. Therefore, because of the consistent behavior of the Coulomb Law for charges at a variety of separation distances, there is, in effect, a single constant angle of deviation that accommodates those of the divergent rays that enter into the effect. There must be some such angle which is essentially the same for all cases until the charges are close enough that the radial divergence affects the result. That angle is termed α_{max} in this development.

In terms of Figure A-1-2, for the case of charges near to each other, α_{max} must accommodate both β and α. When the charges are far apart β is essentially zero so that $\alpha_{max} = \alpha$. But, the maximum angle, α_{max}, all of which is available to α when the ray source is distant, must, when the ray source is near, account first for removing any ray divergence, β, with any remaining balance left for α. Therefore

(A-1-16) $\quad \alpha + \beta = \alpha_{max}$

(The quantity ρ is needed in order for the concept of α_{max} to have meaning; the angle is pointless without defining where it acts. For charges that are far apart α_{max} and ρ are of no significance. When near effects are operating ρ is at $r/2$, half-way between the charges. The concept of ρ is only included here for the initial purpose of presenting in the above Figure A-1-2 the comparison of the near and distant cases.)

The *Reduction Factor* depends upon the reduction of d (of Figure A-1-2) relative to d_{max}, that is the ratio d/d_{max} which quantity is developed as follows.

The angles α, α_{max}, and β are so small that their respective tangents equal their respective angles. Therefore, from the figure

(A-1-17)
$$\mathrm{Tan}[\alpha_{max}] = \alpha_{max} = \frac{d_{max}}{\rho}$$
$$\mathrm{Tan}[\beta_{max}] = \beta_{max} = \frac{d_{max}}{r - \rho}$$
$$\mathrm{Tan}[\alpha] = \alpha = \frac{d}{\rho}$$
$$\mathrm{Tan}[\beta] = \beta = \frac{d}{r - \rho}$$

From which
$$\alpha = \frac{d}{d_{max}} \cdot \alpha_{max}$$
$$\beta = \frac{d}{d_{max}} \cdot \beta_{max}$$

Then, substituting the above results into Equation A-1-16 the following is obtained.

A-1-18 $\quad \alpha_{max} = \alpha + \beta$
$$= \frac{d}{d_{max}} \cdot \alpha_{max} + \frac{d}{d_{max}} \cdot \beta_{max}$$

From which
$$\frac{d}{d_{max}} = \frac{\alpha_{max}}{\alpha_{max} + \beta_{max}}$$

However, α_{max} is a constant quantity (from the consistent Coulomb behavior when the charges are far apart) as is d_{max}.

(A-1-19) $\quad \alpha_{max} = [\text{A Constant}] \cdot d_{max} \equiv \chi \cdot d_{max}$

Substituting for α_{max} of Equation A-1-18 with Equation A-1-19 and for β_{max} of Equation A-1-18 with β_{max} of Equation A-1-17 the *Reduction Factor* sought for Equation A-1-15 is obtained. It is the d/d_{max} of Equation A-1-20, below.

$$(A-1-20) \quad \left[\begin{array}{c}\text{Reduction}\\ \text{Factor}\end{array}\right] = \frac{d}{d_{max}} = \frac{\chi \cdot d_{max}}{\chi \cdot d_{max} + \frac{d_{max}}{r-\rho}}$$

$$= \frac{1}{1 + \frac{1}{\chi \cdot [r-\rho]}}$$

This *Reduction Factor* effect is also the cause of the *Lamb Shift*. The Lamb Shift is an extremely slight shifting to higher energy of the inner orbital energy levels of Hydrogen [Coulomb interaction at close separation as analyzed here]. That is, the Lamb Shift is greater as r is smaller. For that reason, it produces a detectable affect principally on the electrons of the inner orbital shells [$n = 1$ or $n = 2$].

The form of the effect is depicted graphically in Figure A-1-3, below.

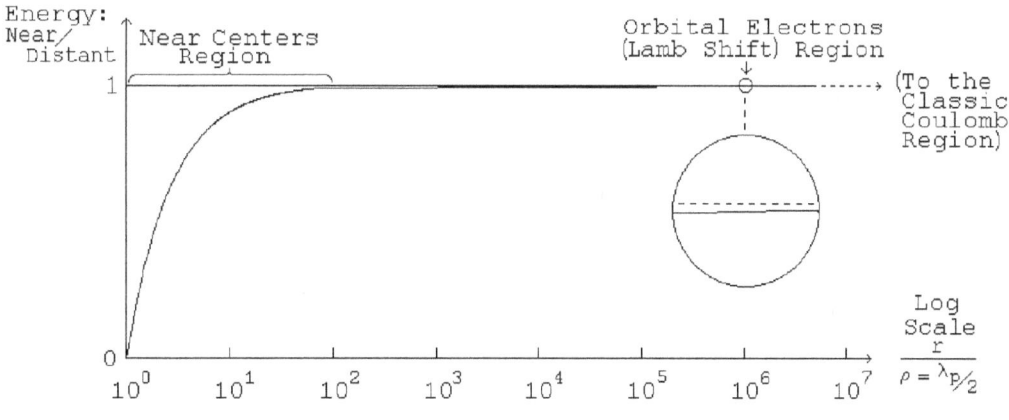

Figure A-1-3
Coulomb Effect <u>Reduction Factor</u> When Charges Are Near to Each Other

The Lamb Shift was attributed to "radiative coupling of the electron to the zero point fluctuation of the vacuum". What that means in plain language is as follows. Heisenberg showed that measurement precision is limited because the information extraction process must change the datum while measuring it. 20th Century physics has questionably extended that to the attribution of a real uncertainty, not merely one of measurement limitation. Then, the zero of the vacuum would also not be precisely zero but a fluctuation in the Heisenberg uncertainty amount about zero. The Lamb shift was attributed to orbital electron interaction with that fluctuation.

The Lamb Shift, is actually caused by the reduction in the negative potential energy due to the orbital electron being near enough to the nucleus that the full Coulomb effect, as when the incoming wave is plane, is slightly reduced as developed above. There being at small values of r marginally less Coulomb attraction, the energy pit in which the electron resides is less deep, which means that its energy is somewhat more than would otherwise be the case. The amount of the effect decreases with increasing r because the reduction in the Coulomb effect decreases as r increases.

The Lamb Shift occurs at much larger values of r (electron orbit radii that are on the order of $r = 10^{-10}$ m) than the quite small value of r at which the neutron forms from the combining proton and electron (on the order of $r = 10^{-15}$ m). Nevertheless, the Lamb Shift can be used for an approximate calibration of the above *Reduction Factor*. The Lamb Shift is depicted in Figure A-1-4, below.

The shift is stated in terms of the wave number (reciprocal wavelength) because the Rydberg expression for the spectral lines is in terms of wave numbers. The amount of the *Balmer Â* shift is 0.033 cm^{-1}. That occurs at the $n = 2$ level where the overall level itself has the term value the Rydberg constant divided by n^2. The fractional shift is then as follows.

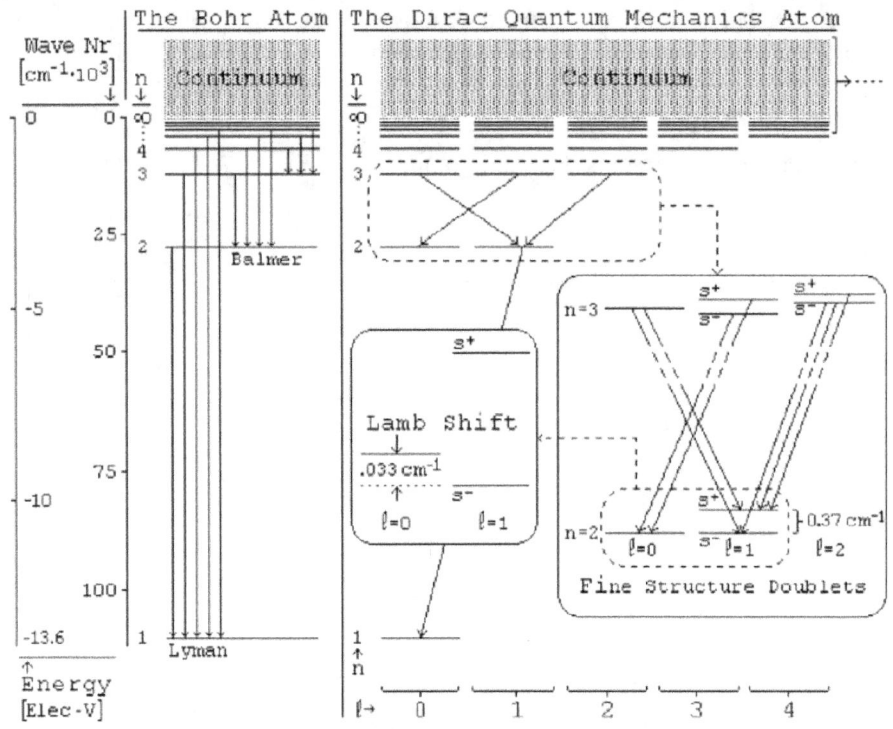

Figure A-1-4
Hydrogen Spectra and the Lamb Shift

(A-1-21) ΔE = Shift = 0.033 cm^{-1} [n=2 Balmer Â shift]

$$E = \text{Total Wave Number} = \frac{Ry}{n^2} = \frac{109,737.31534}{4} = 27,434.3 \text{ cm}^{-1}$$

$$\text{Fractional Shift} = \frac{\Delta E}{E} = \frac{0.033}{27,434.3}$$

$$= 1.2 \cdot 10^{-6} \quad [\text{dimensionless ratio}]$$

The above *Fractional Shift* is the fractional energy change to the "normal" Coulomb potential energy due to the effect of the two charges being near to each other. The *Reduction Factor* as used in this analysis, Equation *A-1-15*, is the net energy after that change, *[1 - the above Fractional Shift]* as follows.

$$\text{(A-1-22)} \quad \begin{bmatrix} \text{Reduction} \\ \text{Factor} \end{bmatrix} = [1 - \text{Fractional Shift}]$$

$$= 1 - 1.2 \cdot 10^{-6}$$

$$= 0.999,998,80$$

$$\frac{1}{1 + \dfrac{1}{\chi \cdot [r - \rho]}} = 0.999,998,80$$

The radius of the $n = 2$ orbit of Hydrogen is $r = 2.1190152 \cdot 10^{-10}$ m. The ρ in the Reduction Factor formula is negligible in the case of the Lamb Shift where $r \approx 10^5 \cdot \rho$ and the precision of the Lamb Shift datum is only two significant digits. Equation A-1-20 can then be solved for the value of χ as follows.

$$\text{(A-1-23)} \quad \chi = \frac{\text{Reduction Factor}}{r \cdot [1 - \text{Reduction Factor}]}$$

$$= 3.9 \cdot 10^{-15}$$

The general formulation for the Reduction Factor is, then, the expression of Equation A-1-18 with the Equation A-1-22 value of χ substituted and $\rho = r/2$. The expression for the potential energy as the proton and the electron approach each other to form a neutron is then Equation A-1-15 with that Reduction Factor substituted. That expression can then be solved for r, the $r_{separation}$ with the following result.

(A-1-24) $r_{separation} = 1.3 \cdot 10^{-15}$ meters

The precision of this result is limited to the two significant digits of the Lamb Shift datum. Nevertheless, it is quite close to the wavelength of the proton oscillation in the neutron per Equation A-1 12, $\lambda_p = 1.321,408,96 \cdot 10^{-15}$ meters.

Alternatively, if $r_{separation}$ is set at λ_p the resulting value for χ can be calculated and from that the value of ΔE, the Lamb Shift. That calculation gives a Lamb Shift of $.033,611,416$ cm^{-1} compared to the actual datum of $.033$ cm^{-1}.

Two conclusions result from these calculations.

First:
The cause of the Lamb Shift is the change in the magnitude of the Coulomb effect when the two charges are near to each other not the "radiative coupling of the electron to the zero point fluctuation of the vacuum".

Second:
The neutron is the combination of a proton and an electron exactly as if each brings to the union its mass equivalent of its escape velocity kinetic energy from the other, the boundary at which the two combine being the wavelength of the proton oscillation, the resulting neutron oscillation being as Figure A-1-1(b) and Equation A-1-1 with f_p and f_e being the frequency equivalents of the masses of Equations A-1-10 and A-1-9 respectively.

Appendix B

The Limitation of the Original Envelopes

This is to show how the otherwise infinite string of envelopes to the original oscillation at the start of the universe was subject to a finite limitation. By "finite limitation" is meant that in the vicinity of the cut-off number of envelopes, N_0, the amplitude of each of the further successive envelopes being imposed on the original $U(t)$, Equation 3-5 was successively significantly less than its immediate predecessor and the rate of that amplitude decrease increased sharply with further envelopes – there was a sharp cut-off of amplitude.

After a moderate number of such cut-off region envelopes the amplitude of any further envelopes becomes infinitesimal. While such infinitesimal (and still continuing to become ever more infinitesimal) envelopes theoretically go on to an infinite number of them, the result is equivalent to the convergence to a finite value of a mathematical infinite series such as, for example that of the cosine. The envelopes cut-off is a result of the mathematics of $U(t)$.

The key to that behavior is to be found in Table B-1, below, the expansion of the $Cos^n(x)$ function. The "Cosmic Egg" expression, Equation 3-5, repeated below

$$(3\text{-}5) \quad U(t) = \pm U_0 \cdot \left[1 - \cos\left[2\pi f_{env} \cdot t\right]\right]^{N_0} \cdot \left[1 - \cos\left[2\pi f_{wve} \cdot t\right]\right]$$

contains the factor

$$(B\text{-}1) \quad \cos^{N_0}\left[2\pi(f_{env})t\right]$$

which creates the set of envelopes to the original oscillation. The expansion of the cosine raised to the power of its N_0 exponent behaves according to the pattern illustrated in Table B-1, below. Analysis of the patterns in the coefficients of the individual terms of the $Cos^n(x)$ expansion discloses a pattern related to the binomial expansion as demonstrated in the table.

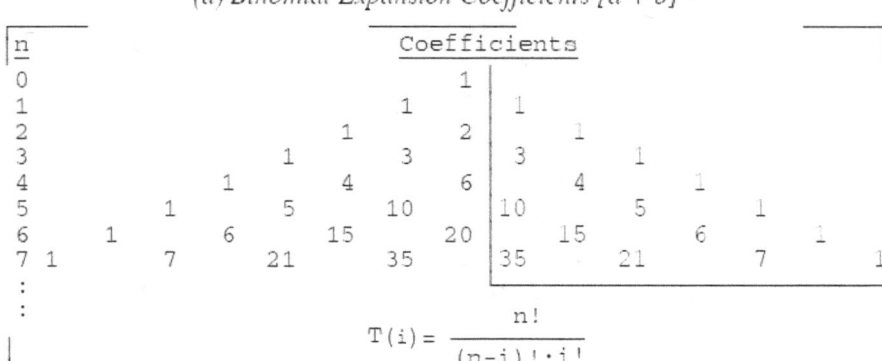

(a) Binomial Expansion Coefficients $[a + b]^n$

n					Coefficients				
0					1				
1				1		1			
2				1	2	1			
3			1	3		3	1		
4			1	4	6	4	1		
5		1	5	10		10	5	1	
6	1	6		15	20	15		6	1
7	1	7	21		35	35	21	7	1
⋮									

$$T(i) = \frac{n!}{(n-i)! \cdot i!}$$

(b) $Cos^n(x)$ Expansion Coefficients

n			Coefficients						
	Times Cos(*), * =	0x	1x	2x	3x	4x	5x	6x	7x
0		1							
1		-	1						
2		1	-	1					
3		-	3	-	1				
4		3	-	4	-	1			
5		-	10	-	5	-	1		
6		10	-	15	-	6	-	1	
7		-	35	-	21	-	7	-	1
⋮									

$$T(i) = \frac{n!}{(n-i)! \cdot i!}$$

Table B-1

Clearly, with the exception of the constant term (where, in the table, * = *0x*) the other terms of the expansion of $Cos^n(x)$ have the same coefficients as the corresponding terms of the binomial expansion. The formula for the binomial expansion can thus be used to obtain the coefficients for any value of *n* in the expansion of $Cos^n(x)$. in the present case for any value of N_0 in the expansion of the *U(t)* factor $Cos^{N_0}[2\pi(f_{env})t]$

The cut-off occurs around the value of N_0 regardless of what that value is. Therefore the value of N_0 is not important. Nevertheless it is of interest that various attempts to estimate it give values around 10^{85}.

$N_0 = 10^{85}$ is the *n* of the formula. It is not practicable and most likely not possible to calculate all of the coefficients of the cosine expansion of the envelopes for 10^{85} envelopes. On the other hand, it is not unreasonable to calculate the *85* cases corresponding to the frequency multiples of the expansion: 10^1, 10^2, 10^3, ⋯ 10^{85}.

Figure B-1, below, is a plot of the relative magnitude of the successive coefficients of the various frequency multiples $(1 \cdot x, 3 \cdot x, \cdots 10^{85} \cdot x)$, in the expansion of $Cos^n(x)$ for $n = N_0 = 10^{85}$. The plot indicates a sharp cut-off, an attenuation of the higher frequencies. Figure B-1(a) uses a linear horizontal axis and shows the cut-off in detail. Figure B-1(b) uses a logarithmic horizontal scale to better present the tremendous range in frequency multiples from 1 to 10^{85}. It shows that the cut-off is quite sharp and drastic.

This cut-off is merely the action of the mathematics of $cos^n(x)$.

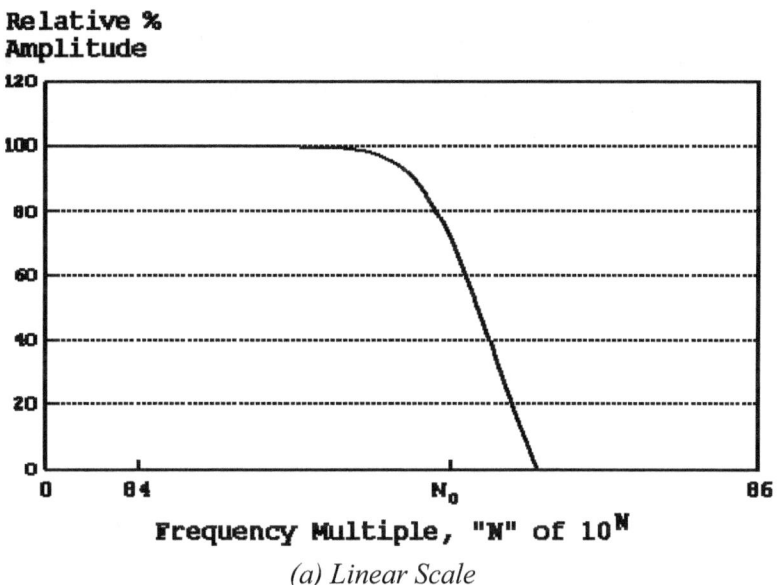

(a) Linear Scale

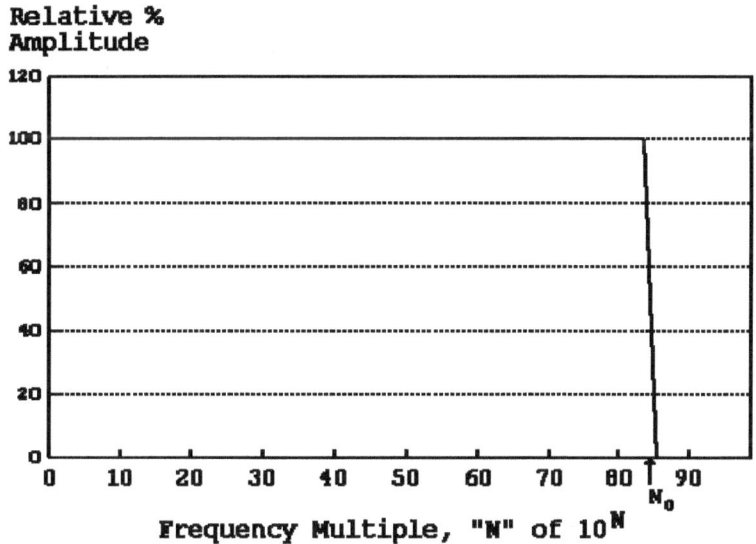

(b) Logarithmic Scale

Figure B-1
The $Cos^n(x)$ Limitation of the "Cosmic Egg

Appendix C

Derivation of Coulomb's Law

The fundamentals of what is known about the actions of electric charge can be summarized as follows.

- Electric charges exist in two different forms, termed the *sign* or *polarity* of the charge, "positive" or "negative".

- The charges exert a force on other electric charges.

- The force attracts or repels for charges of opposite or like signs.

- The force is inversely proportional to the square of the separation distance of the charges and directly proportional to the amounts of the charges.

- The effect extends throughout space.

- The charges only exist as a component effect of particles having mass which particles have been shown in the preceding sections to be *Spherical-Centers-of-Oscillation*.

The details of this behavior have been thoroughly worked out in terms of mathematics which describe the location, amount, and direction of the effect. The physical constants needed to give correct quantitative results have been well determined.

Each electrically forc**ing** particle *[Spherical-Center-of-Oscillation]* must communicate to each electrically forc**ed** particle *[Spherical-Center-of-Oscillation]* the direction from the forc**ing** particle to the forc**ed** one [for same signs repulsion], the direction from the forc**ed** particle to the forc**ing** one [for opposite signs attraction] and the magnitude and sign of the forc**ing** particle's charge. That task is assigned by contemporary physics' theory to an *electric field*, a vector field that is an assignment of a direction of action and its magnitude to each point in a region of space.

However, that designation of the field, while facilitating the description of the action fails to explain the cause, the mechanism of the field and thus fails to explain or account for the action at issue. It also fails to account for the time delay, due to the limitation of the speed of light, that must exist between a change at the forc**ing** particle and its effect at the forc**ed** particle

A flow, flowing at the speed of light, continuously, carrying the direction and magnitude information, spherically outward, from every electrically acting *Spherical-Center-of-Oscillation* to every other such *Spherical-Center-of-Oscillation*, from every charge to every other, is required. That *Propagated Outward Flow* was introduced and described in Section 3.

How the Charges and Their Flow Repel and Attract

The effect of an individual wave of that *Propagated Outward Flow* encountering another *Spherical-Center-of-Oscillation* is the delivery of a train of impulses to the center, Figure C-1, each an amount of momentum. That is

(C-1)   impulse = force·time = mass·velocity = momentum

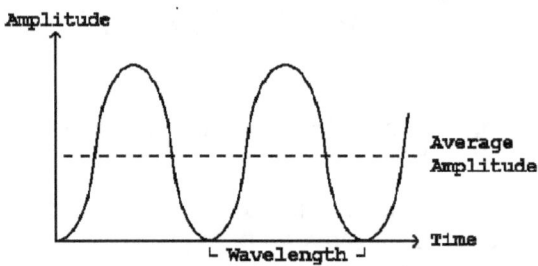

Figure C-1
The +U Wave of the Propagated Outward Flow from a +U Spherical Center-of-Oscillation

The wave as it is propagated by its source *Spherical-Center-of-Oscillation*, carries potential impulse, "potential" because it is not realized in an effect until an encounter with another *Spherical-Center-of-Oscillation* occurs. The amount of potential impulse in the wave is, of course, proportional to the amplitude of the wave. It is that amplitude, which decreases as the square of the distance from the source *Spherical-Center-of-Oscillation* because it becomes spread over a greater area.

The overall stream of waves carries the potential impulse of one wave times the repetition rate, the frequency, of the waves. The potential status of the wave's impulse is exactly the same status as that of electric field (which it, in fact, is) where electric field is potential force and not realized as actual force until it interacts with another charge.

A *Spherical-Center-of-Oscillation* propagating a *+U Wave Propagated Outward Flow* experiences an equal *Spherical-Center-of-Oscillation* magnitude, opposite direction reaction to the radially outgoing train of impulses as if the *Spherical-Center-of-Oscillation* were under spherical compression, Figure C-2. However, that is to no net effect because of its spherical symmetry.

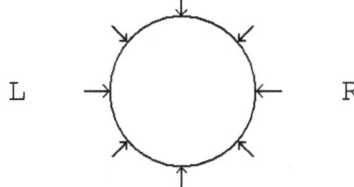

Figure C-2
The +U Spherical-Center-of-Oscillation's Reaction Back On Itself by Its Outward Flow

The train of impulses of Figure C-1 encountering the *Spherical-Center-of-Oscillation* of Figure C-2 on its left side [L] adds additional momentum to the reaction directed to the right. That being now greater than the opposing reaction to the left on the right side [R], there is now a net momentum increment to the right, a repelling action of one positive charge on another.

If the *Spherical-Center-of-Oscillation* of Figure C-2 were a -*U* center the effect would be reversed. The train of +*U* impulses of Figure C-1 encountering the center of Figure C-2 as a -*U* center on its left side [L] subtract from or cancel part of its reaction directed to the right. That being smaller than the opposing reaction to the left on the right side [R], there is a net momentum increment to the left. The effect is an attracting action of a positive charge on a negative one.

The effects and action are exactly analogous for the two other cases of a -*U* *Spherical-Center-of-Oscillation*'s train of -*U* impulses encountering another -*U* *Spherical-Center-of-Oscillation* or a +*U* one.

It is important to observe that the direction of momentum actions is the direction of the *Propagated Outward Flow* transmitting them whereas the sign or polarity +*U* or -*U* pertains back to the origin of the oscillations that started the "Big Bang" – a pair of exact opposites necessary to maintain conservation.

Having obtained from the *Spherical-Centers-of-Oscillation* and their *Propagated Outward Flows* the directions and polarities of Coulomb's Law it is now necessary to definitively quantify the action.

NEWTON'S LAW AND CENTERS & WAVES – "RESPONSIVENESS"

Newton's Second Law and as restated by inversion are:

(C-2)   Force = Mass · Acceleration

Acceleration Resulting = Force Applied × $1/_{Mass}$

which translates in terms of the waves of *Propagated Outward Flow*s and *Spherical-Centers-of-Oscillation*s into

(C-3) $\begin{bmatrix} \text{Acceleration} \\ \text{Resulting} \end{bmatrix} = \begin{bmatrix} \text{Wave} \\ \text{Impulse} \end{bmatrix} \cdot \begin{bmatrix} \text{Responsiveness} \\ \text{of the Center} \end{bmatrix}$

or, more succinctly,

Acceleration = Wave × Responsiveness.

Of the total wave traveling outward from the source *Spherical-Center-of-Oscillation*, the only part that interacts with another *Spherical-Center-of-Oscillation* is the part intercepted by the encountered center. The *Spherical-Center-of-Oscillation* intercepting the larger portion of incoming wave receives the greater impulse, the greater momentum change. Thus center responsiveness depends on the encountered center's cross-section target for interception of *Propagated Outward Flow* waves.

(This analysis assumes that the part of the wave intercepted by the encountered center is a flat wave front. The non-plane wave case, for small separation distances, is in most cases of negligible effect except the slight "Lamb Shift" treated in Appendix A-1, *The Neutron*, Likewise, because δ, the encountered particle's core radius, is so minute the target can be deemed flat)

A *Spherical-Center-of-Oscillation* of smaller cross-section is of greater mass (lesser responsiveness). The encountered center being a spherical oscillation the cross-section is the area of a circle perpendicular to the direction of travel of the incoming wave front as it encounters the center. That area is <u>proportional</u> to π times the square of the center's wavelength.

This yields the first factor in *Spherical-Center-of-Oscillation* responsiveness,

(C-4) $\quad$ Cross-section $\propto \pi \cdot \lambda_c^2 = K_{cs} \cdot \lambda_c^2$

(C-5) $\quad \begin{bmatrix} \text{respon-} \\ \text{siveness} \end{bmatrix} \propto$ [Factor 1]·[Factor 2]·[Factor 3]

$\qquad\qquad\qquad = [K_{cs} \cdot \lambda_c^2] \cdot [\quad " \quad] \cdot [\quad " \quad]$

where: K_{cs} = a constant for the proportionality
$\qquad\quad \lambda_c$ = the encountered center oscillation wavelength

The incoming wave must be expressed in terms of "Wave Impulse per Unit Area" so that multiplied by the cross-section area at the encountered *Spherical-Center-of-Oscillation* the units of area are cancelled and the resulting quantity is wave impulse.

(C-6) $\quad$ "Wave" = $\dfrac{\text{Total Source Center Propagated Wave}}{\text{Total Spherical Area of Source Wave at Distance Encountered Center is from Source}}$

$\qquad\qquad\quad$ = Wave Impulse per Unit Area

Factor 2 in the responsiveness, Equation C-5 is the <u>effective</u> amplitude of the *Spherical-Center-of-Oscillation*'s oscillation. A range of possible interactions can occur because the source and encountered center frequencies may differ. The extremes and mean of the range of encounters follow.

(1) Frequency$_{source}$ << Frequency$_{encountered}$

The encountered center goes through all of its amplitude values many times while one source wave arrives. Its effective amplitude is its average amplitude.

(2) Frequency$_{source}$ >> Frequency$_{encountered}$

The source center goes through all of its amplitude values many times while the encountered does once. Its effective amplitude is its average amplitude.

(3) Frequency$_{source}$ = Frequency$_{encountered}$

The interaction takes place over exactly one cycle and the effective amplitude is, again, the average.

In real matter, not the idealized model of one source and one encountered center, every *Spherical-Center-of-Oscillation* is constantly "bombarded" by various waves from a variety of directions at a variety of frequencies and phases due to the immense number of *Spherical-Centers-of-Oscillation* making up ordinary matter. The relative frequency and the phase of the wave and the encountered center have no effect on the large scale result from the interaction. Thus *Factor 2* is not a variable quantity but merely the average amplitude of the encountered center, which is designated U_c.

However, the absolute frequency of the encountered *Spherical-Center-of-Oscillation* is *Factor 3* in the formula for responsiveness. Just as the incoming wave repetition rate affects the amount of force that the wave can deliver to the encountered center, so the encountered center repetition rate affects that center's response to the wave. While the wave is encountering the center, each cycle of the encountered center's

C – DERIVATION OF COULOMB'S LAW

oscillation is acted on by the wave. (This is most easily visualized if the frequency of the encountered center is much larger than that of the wave, but it applies in any case.)

Thus *Factor 3* is encountered center repetition rate. [For a center at rest the "rep rate" is the oscillation frequency but for a center in motion its velocity is a factor in the "rep rate" along with its oscillation frequency.

Then Equation *C-5* becomes

(C-7)   responsiveness ∝ [cross-section]·[amplitude]·[rep rate]

$$= [K_{cs} \cdot \lambda_c^2] \cdot [U_c] \cdot [f_c]$$

$$= K_{cs} \cdot \lambda_c \cdot U_c \cdot c \qquad \text{[Using } c = f \cdot \lambda\text{]}$$

where: K_{cs} = a constant for the proportionality
λ_c = the encountered center oscillation wavelength
U_c = its amplitude, and
f_c = its frequency.

PRECISE FORMULATION OF COULOMB'S LAW

The treatment here is of one single unit charge, $\pm U_c \cdot [1 - \cos(2\pi \cdot f \cdot t)]$, interacting with another such single unit charge, one simple basic *Spherical-Center-of-Oscillation* interacting with another.

The Encountered Center Charge Q_e and Its Amplitude U_c

In the traditional formulation of Newton's Law Equation *C-8*

(C-8)   Force = mass·acceleration

and for the case that is now being considered, that in which the force results from the electrostatic interaction between two charges in accordance with Coulomb's Law, Equation *C-9*,

(C-9) $$\text{Force} = \frac{\text{Charge} \cdot \text{Charge}}{\text{Separation Distance}^2}$$

both of the charges enter into the relationship in the *Force* part, the *Mass* part of the relationship being like an inert characteristic of the substance.

In this *Centers-of-Oscillation* formulation Equation *C-2*, repeated here

(C-2) $$\begin{bmatrix} \text{Acceleration} \\ \text{Resulting} \end{bmatrix} = \begin{bmatrix} \text{Wave} \\ \text{Impulse} \end{bmatrix} \cdot \begin{bmatrix} \text{Responsiveness} \\ \text{of the Center} \end{bmatrix}$$

or, more succinctly,

Acceleration = Wave × Responsiveness

the amplitude of the oscillation, U_c for the center, U_w for the wave, the role of which corresponds to that of traditional charge, Q, enters into the formulation differently from the traditional conception. The source *Spherical-Center-of-Oscillation*'s amplitude is a factor in the Wave and the encountered *Spherical-Center-of-Oscillation*'s amplitude is a factor in the Responsiveness.

Figure C-3 on the following page compares the two.

Field and Wave, not Force and Wave, correspond. Each is the unrealized potential that becomes action via interaction with an encountered charge / center. Therefore the [Charge ÷ Mass] of the left half of Figure C-3 is the same as the Responsiveness of the right half of the figure.

$$\begin{array}{ll}
\text{Traditional} & \text{Centers - of - Oscillation} \\[6pt]
\text{Acceleration} = \text{Force} \times \left[\dfrac{1}{\text{Mass}}\right] & \text{Acceleration} = \left[\begin{array}{c}\text{Wave}\\ \text{Impulse}\end{array}\right] \times \text{Responseness} \\[10pt]
= \left[\dfrac{Q \cdot Q}{d^2}\right] \times \left[\dfrac{1}{\text{Mass}}\right] & \\[10pt]
= \left[\dfrac{Q_s}{d^2}\right] \times \left[\dfrac{Q_e}{\text{Mass}}\right] & \\[10pt]
= \left[\begin{array}{c}\text{Electric}\\ \text{Field at } d^2\end{array}\right] \times \left[\dfrac{Q_e}{\text{Mass}}\right] & = \left[\begin{array}{c}\text{Wave}\\ \text{Impulse}\end{array}\right] \times \left[K_{cs} \cdot \lambda_c \cdot U_c \cdot c\right]
\end{array}$$

Figure C-3

Therefore

(C-10) $$\frac{Q_e}{m_e} = K_{cs} \cdot \lambda_c \cdot U_c \cdot c$$

from which

(C-11) $$Q_e = \frac{h}{\lambda_c \cdot c}[K_{cs} \lambda_c \cdot U_c \cdot c] \qquad [\text{Using } mc^2 = h \cdot f]$$
$$= h \cdot K_{cs} \cdot U_c$$

which relates the charge of the encountered *Spherical-Center-of-Oscillation* to it's amplitude, and is a simple direct proportionality because h and K_{cs} are constants.

The Source Center Charge Q_s and Its Oscillation Amplitude U_c

If time could be stopped so that the waves from the source center were frozen in whatever position that they had in space, then the spherical waves as propagated by a *Spherical-Center-of-Oscillation* would appear as a series of nested shells, each of a successively greater radius, R, the radius being

(C-12) $R_w = n \cdot \lambda_w$

where: n = 1, 2, 3 ... for the successive shells
λ_w = the wavelength of the waves

and the thickness of each shell is the wavelength, λ_w. One such shell is depicted two-dimensionally in Figure C-4, below.

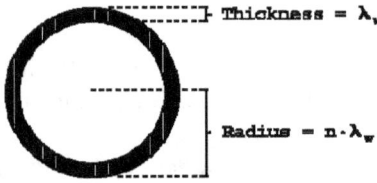

Figure C-4

A cross-sectional view of this wave in space, that is a graph of its amplitude variation along a radius while traversing the thickness, is depicted in Figure C-5, below, where it is clear that the area under the curve of amplitude variation is equal to $U_w \cdot \lambda_w$.

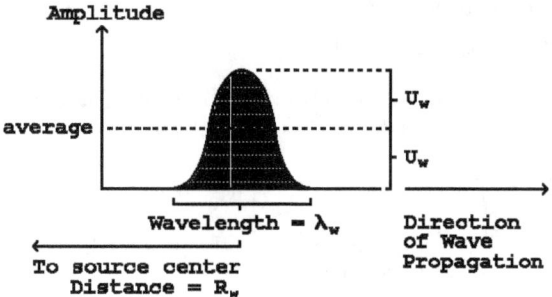

Figure C-5

The potential impulse in one complete spherical shell, one wave cycle, is the shell cross-section, $U_w \cdot \lambda_w$, multiplied by the spherical surface area of the shell, $4\pi \cdot R_w$.

(C-13) [a cycle of wave impulse] = $[U_w \cdot \lambda_w] \cdot [4\pi \cdot R_w]$

But, the wave amplitude, U_w, is the *Spherical-Center-of-Oscillation*'s amplitude, U_c, divided by the area of the wave's spherical shell at R_w and $\lambda_w = \lambda_c$ so that

(C-14) [a cycle of wave impulse] = $U_c \cdot \lambda_c$

The *Wave* of Figure C-3 is the Equation C-14 single *[a cycle of wave impulse]* multiplied by the repetition rate, the frequency, $f_w = f_c$, so that the wave, of Figure C-3 is

(C-15) Wave = $[U_c \cdot \lambda_c] \cdot f_c = U_c \cdot c = Q_s$,

which relates the field of the source *Spherical-Center-of-Oscillation* to that center's oscillation amplitude and, therefore, relates the charge of the source center to its amplitude.

Recognizing that every *Spherical-Center-of-Oscillation* is always in both source and encountered roles, then setting Equation C-11 equal to Equation C-14 the following is obtained.

(C-16) $Q_e = Q_s$

$h \cdot K_{cs} \cdot U_c = U_c \cdot c$

therefore

$Q = U \cdot c$ and $K_{cs} = c/h$

Two Such Charges Interact Electrostatically As Follows

(1) The total potential force in the wave series as propagated by the source *Spherical-Center-of-Oscillation*s is (from Equation C-15)

(C-15) $U_c \cdot c$

(2) The total wave series potential force per unit area of wave front at the encountered *Spherical-Center-of-Oscillation* is the quantity of step (1) divided by the spherical surface at the encountered center.

(C-17) $\dfrac{U_c \cdot c}{4\pi \cdot R^2}$

(3) The responsiveness of the encountered *Spherical-Center-of-Oscillation* is (Equation C-7)

(C-7)   Responsiveness = $K_{cs} \cdot \lambda_c \cdot U_c \cdot c$

(4) The resulting acceleration is, therefore (substituting steps (2) and (3), above, into Equation C-3 per Equation C-6)

(C-18)
$$\text{Acceleration} = \begin{bmatrix} \text{Wave Potential} \\ \text{Impulse per Unit} \\ \text{Area at the En-} \\ \text{countered Center} \end{bmatrix} \cdot \begin{bmatrix} \text{Responsiveness} \\ \text{of the} \\ \text{Encountered} \\ \text{Center} \end{bmatrix}$$

$$= \frac{U_c \cdot c}{4\pi \cdot R^2} \cdot K_{cs} \lambda_c U_c c$$

(5) The mass of the encountered *Spherical-Center-of-Oscillation* (from $m \cdot c^2 = h \cdot f$) is

(C-19) $m = \dfrac{h}{c \cdot \lambda_c}$

(6) The force is, then (substituting steps (4) and (5), above into Equation C-2)

(C-20)   Force = Mass × Acceleration

$$= \left[\frac{h}{c \lambda_c}\right] \cdot \left[\frac{U_c \cdot c}{4\pi \cdot R^2}\right] \cdot K_{cs} \lambda_c \cdot U_c \cdot c$$

$$= \frac{[U_c \cdot c] \cdot [h \cdot K_{cs} \cdot U_c \cdot c]}{4\pi \cdot R^2}$$

and substituting per Equations C-11 and C-15 yields the result

(C-21) $\text{Force} = \dfrac{Q_s \cdot Q_e}{4\pi \cdot R^2}$

which is Coulomb's law as it naturally occurs.

If a constant of proportionality, k, is introduced to accommodate choice of the units of charge, and the 4π is absorbed into that new constant, then the result (using q for charge since the added constant requires an accordingly different variable) is

(C-22) $\text{Force} = k \cdot \dfrac{q_s \cdot q_e}{R^2}$ $k = {}^{1}/_{4\pi\varepsilon_0}$

which is Coulomb's Law as originally formulated.

[* See on the following page the analysis

"Understanding:

The Units of Charge and of Coulomb's Law"]

ANALYSIS

Understanding:
The Units of Charge and of Coulomb's Law

Properly stated, the law of electrostatic interaction between two charges, called Coulomb's Law, is

> "Given two electric charges separated in space by some distance, the magnitude of the force exerted by each of the charges on the other is directly proportional to the product of the charges and inversely proportional to the square of the distance between them."

In symbols this is

(C-A-1)
$$F = k \cdot \frac{Q_1 \cdot Q_2}{R^2}$$

where k = the constant of the proportionality.

Unfortunately the manner in which the law was originally formulated and other complications led to various systems of units.

It is desirable for simplicity that the units for the quantities in such laws be so as to have the constant of proportionality, k, be unity. Then the constant of proportionality can be omitted and the statement of the law involves only the actual variables pertinent to the law.

There are many examples of physical laws in which this was accomplished:

```
Force = Mass × Acceleration
(not k × Mass × Acceleration)

Voltage = Current × Resistance
(not k × Current × Resistance)
```

and so forth.

Of course, what is desired is that this be done ($k=1$) successfully for all systems of units that might be used. Commonly encountered systems of units are:

```
(1) cgs ≡ length in centimeters (cm)
          mass in grams (gm)
          time in seconds (sec)
```

with other units following accordingly as prescribed by the physical laws involved, for example:

$$\text{force} = \text{mass} \times \text{acceleration}$$
$$= \text{mass} \times \text{length}/\text{time}^2$$
$$= \text{gm} \times \text{cm}/\text{sec}^2$$
$$\equiv \text{dyne}$$

(2) MKS ≡ length in meters (m)
      mass in kilograms (kg)
      time in seconds (sec)

$$\text{force} = \text{kg} \times \text{m}/\text{sec}^2$$
$$\equiv \text{newton}$$

It would appear then that one need merely rearrange Coulomb's Law so that it can be used to define the units of charge as

(C-A-2) $$Q^2 = \frac{F \times R^2}{k}$$ [C-A-1 rearranged, and since this is only for units, $Q_1 = Q_2 = Q$]

and an orderly, simple arrangement of units would result. But, unfortunately it does not. From equation *C-A-2* the units of *Q* are as follows.

(C-A-3) $Q = [F \times R^2]^{1/2}$ [k is dimensionless for finding the natural units of Q]

$$= [\text{force} \times \text{length}^2]^{1/2}$$

$$= [(\text{mass} \times \text{acceleration}) \times \text{length}^2]^{1/2}$$

$$= \left[\text{mass} \times \frac{\text{length}}{\text{time}^2} \times \text{length}^2\right]^{1/2}$$

$$= \left[\frac{\text{mass} \times \text{length}^3}{\text{time}^2}\right]^{1/2}$$

The following table, Figure C-A-1, indicates the manner in which common units work out in this formulation in different systems of units.

Quantity	cgs Units	MKS Units	Ratio cgs/MKS
length	cm	m	10^2 cm/m
mass	gm	kg	10^3 gm/kg
time	sec	sec	1
velocity	cm/sec	m/sec	10^2 cm/m per sec
force	dyne	newton	10^5 dyne/newton
. . .	. . .	. . .	. . .
charge	$\left[\dfrac{\text{gm} \times \text{cm}^3}{\text{sec}^2}\right]^{1/2}$	$\left[\dfrac{\text{kg} \times \text{m}^3}{\text{sec}^2}\right]^{1/2}$	$10^{9/2} = 10^4 \cdot \sqrt{10}$ cgs units per MKS unit

Figure C-A-1

C – DERIVATION OF COULOMB'S LAW

Most quantities in the table experience value changes by a multiple factor of ten in going from *cgs* to *MKS* units, which is simple. The digits of a quantity remain the same, only the position of the decimal point changes from one system of units to another. But charge will not fit that simple pattern. If for example a certain value of charge in some situation in *cgs* units were to be $4.803 \cdot 10^{-10}$ the corresponding value in *MKS* units would be $1.519 \cdot 10^{-14}$. That is not simple and orderly as desired because the digits of the quantity as well as the power of ten change with the change in units used.

A second complication arises when it is further found that the k in Coulomb's law is not always the same. Its value depends on the nature of the material substance (or lack of it) intervening between the two charges, for example: air, glass, oil, free space (perfect vacuum), etc.

Recognizing, then, that the desirable procedure of choosing the fundamental unit of new quantities so that $k = 1$ is hopeless in this case of charge, k is set up as a constant that is dependent on the intervening material and retained as part of the physical law. The k is designated $1/\varepsilon$ (Greek letter epsilon). For free space the epsilon is designated as ε_0. For one system of units the "natural electrostatic units" (and for the free space condition) $k = 1$ can still be retained as $\varepsilon_0 = 1$.

Coulomb's Law then becomes

(C-A-4)
$$F = \frac{Q_1 \cdot Q_2}{\varepsilon \cdot R^2} \qquad \text{[Anywhere]}$$

and

(C-A-5)
$$F = \frac{Q_1 \cdot Q_2}{\varepsilon_0 \cdot R^2} \qquad \text{[In free space]}.$$

The more or less orderly arrangement in Figure C-A-2, below, results as established in practice, and is as simple and orderly as can be obtained in the circumstances. The "Elemental Charge" in the table is the value of the charge of an electron or a proton, the fundamental electric charge of the universe, in each system of units.

The esu (cgs, natural electrostatic) units is the system in which Coulomb's Law was originally developed (implicitly), where the value of the charge is $4.803 \cdot 10^{-10}$ and $\varepsilon_0 = 1$. The units, abcoulombs, are equal to statcoulombs divided by the velocity of light, c. A *coulomb* is *10 abcoulombs*. The rationalized system of units recognizes the significance of the 4π factor in the law and takes it into the ε_0 rather than the more awkward step of changing the charge to its otherwise value times $1/\sqrt{4\pi}$.

(For a thorough analysis of systems of units see Chapter 3, *Handbook of Engineering Fundamentals*, First Edition, Ovid W. Eshbach, New York, John Riley & Sons, 1947.)

The *Rationalized Meter - Kilogram - Second (MKSR)* system in Figure C-A-2, below, is now established as the standard system of units to be used internationally. The system is now referred to as *SI Units*, that is *Standard International* Units.

System of Units	Elemental Charge	Value of ε_0	Correct Statement of Coulomb's Law For F in Free Space
esu (cgs, natural electrostatic)	$4.803 \cdot 10^{-10}$ statcoulombs (Q)	1	$F = \dfrac{Q_1 \cdot Q_2}{\varepsilon_0 \cdot R^2}$ $= \left[\dfrac{Q_1 \cdot Q_2}{R^2}\right]$
emu (cgs, natural electromagnetic with ε as it "naturally" occurs, see after the Figure)	$1.602 \cdot 10^{-20}$ abcoulombs (q_{abs})	$\dfrac{1}{c^2}$	$F = \dfrac{q_{1abs} \cdot q_{2abs}}{\varepsilon_0 \cdot R^2}$ $= \dfrac{c^2 \cdot q_{1abs} \cdot q_{2abs}}{R^2}$ $= \left[\dfrac{Q_1 \cdot Q_2}{R^2}\right]$
MKS (MKS with an adjustment factor in ε)	$1.602 \cdot 10^{-19}$ coulombs (q)	$\dfrac{10^7}{c^2}$	$F = \dfrac{q_1 \cdot q_2}{\varepsilon_0 \cdot R^2}$ $= \dfrac{c^2 \cdot q_1 \cdot q_2}{10^7 \cdot R^2}$ $\genfrac{}{}{0pt}{}{* \to}{** \to} = \dfrac{(10^4 \cdot \sqrt{10})^2}{10^2} \cdot \dfrac{c^2 \cdot q_{1abs} \cdot q_{2abs}}{10^7 \cdot R^2}$ $= \left[\dfrac{Q_1 \cdot Q_2}{R^2}\right]$
MKSR ("rationalized" MKS)	$1.602 \cdot 10^{-19}$ coulombs (q)	$\dfrac{10^7}{4\pi \cdot c^2}$	$F = \dfrac{q_1 \cdot q_2}{4\pi \cdot \varepsilon_0 \cdot R^2}$ $= \dfrac{c^2 \cdot q_1 \cdot q_2}{10^7 \cdot R^2}$ $= \left[\dfrac{Q_1 \cdot Q_2}{R^2}\right]$

* (This is the ^{cgs}/MKS units ratio applied to each charge.)
** (The ^{abcoulomb}/coulomb ratio applied to each charge.)

Figure C-A-2

Appendix D

The Universal Exponential Decay

Since the "Big Bang" the *Propagated Outward Flow* has been gradually depleting the original supply of *medium* in the core of each *Spherical-Center-of-Oscillation*. That process, an original quantity gradually depleted by flow away of some of the remaining quantity, is an exponential decay.

THE NATURE OF THE DECAY

Of the three fundamental dimensions of length $[L]$, mass $[M]$, and time $[T]$ only length can decay. Time being the independent variable of material reality, whether it decays, varies, or is rigorously constant is beyond our ability to detect. Likewise, mass cannot decay, it being proportional to frequency, the inverse of time.

The dimension that is decaying is length, the $[L]$ dimension in the dimensions of, for example: the Planck Constant, h, $[M \cdot L^2/T]$; the speed of light, c, $[L/T]$; and the Newtonian Gravitational Constant, G, $[L^3/M \cdot T^2]$.

The Evidence

Four independent unrelated phenomena, none of which has an established explanation, have now been extensively observed and a large amount of data substantiating the phenomena have been developed. The phenomena are as follows

In 1933 F. Zwicky reported [1] that the rotational balance of gravitational central attraction and rotational centripetal force in galaxies appeared to be out of balance, that a small additional centrally directed acceleration of unknown source appeared to be needed and to be acting. Numerous galactic rotation curves confirm that there is such an anomalous acceleration present and necessary in all rotating galaxies.

- In 1998 the Pioneer Anomaly was first reported [2]. The anomaly is a small acceleration, centrally directed [toward the Sun], constant, distance independent, and of unknown cause, observed in the tracking of the Pioneer 10 and 11 spacecraft from launch until their near departure from the Solar System.

- In 2008 the Flybys Anomaly was first reported [3]. The anomaly is unaccounted for changes in spacecraft speed, both increases and decreases, for six different spacecraft involved in Earth flybys from December 8, 1990 to August 2, 2005.

- Also in 2008 a previously unknown large scale flow of galaxy clusters all in the same direction toward "the edge" of the observable universe, the Dark Flow anomaly was first reported [4]. The mysterious motion, originally noted in 2008 using the three-year WMAP survey, is now [2010] confirmed by a more comprehensive five-year study [7].

Analysis discloses that the first three have in common the same locally centrally directed, small acceleration that is non-gravitational, distance independent, constant, and unaccounted for. The fourth phenomenon is shown to be fully consistent with that same cause and explanation.

The Centrally Directed Anomalous Acceleration in all Rotating Galaxies

In general, galaxies are rotating systems, a balance of gravitational attraction $[G \cdot M \cdot m / R^2]$ and centripetal force $[m \cdot V^2 / R]$ maintaining the structure. A curve or plot of such rotational velocity, V, versus path radius, R, is termed a Rotation Curve.

When the central mass is far greater than the orbiting masses the dynamics are such that the orbital velocities are inversely proportional to the square root of the radial distance from the center mass $[V = (G \cdot M / R)^{1/2}]$, as for example in our solar system and as illustrated in Figure D-1, below. Such rotational dynamics and rotation curves are referred to as Keplerian.

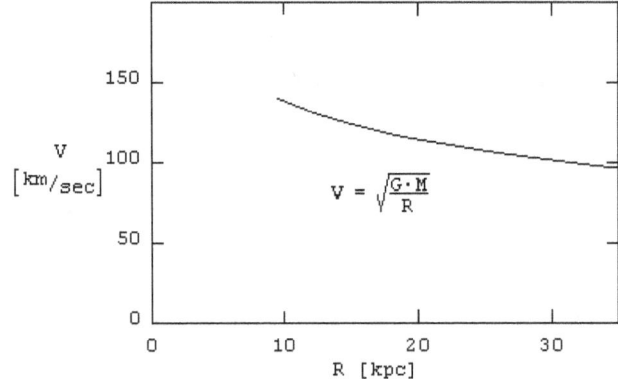

Figure D-1 - A Keplerian Rotation Curve

In the case of a solid sphere of uniform density, ρ, throughout, all parts must move at rotational velocities directly proportional to radius as illustrated in Figure D-2.

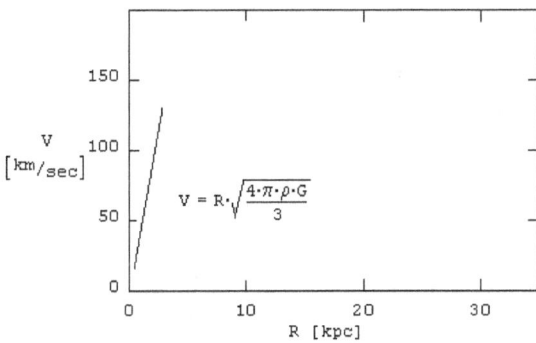

Figure D-2 - The Rotation Curve of a Solid Sphere of Uniform Density

The form of galaxies as we are able to directly observe them is that of a fairly spherical star-dense central core and a transition from that to the much more extensive flat disk of a far smaller density of more widely dispersed stars. The portion of galactic rotation curves that pertains to the dense central core of the galaxy would be expected to exhibit approximately the same velocity-proportional-to-radius form as illustrated for a solid sphere in Figure D-2, above. Likewise, the more dispersed flat disk, minor in mass compared to the dense central core, would be expected to exhibit the Keplerian form of Figure D-1, above. The expected form of galactic rotation curves would be the two combined with a smooth transition between as Figure D-3.

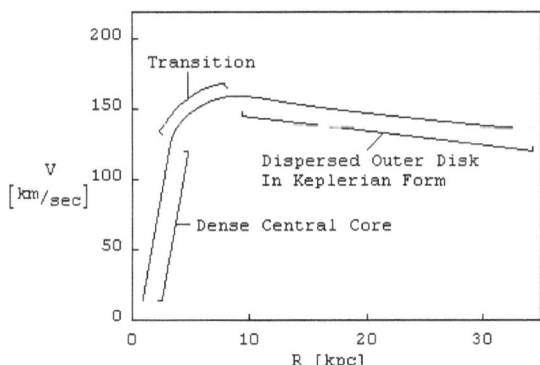

Figure D-3 - The Expected Form of Galactic Rotation Curves

For galaxies that present themselves in an edge view of the thin disk not as their spiral or globular spread in space, it is possible to measure the rotational velocities and obtain a rotation curve. We see one end of the presented flat disk moving toward us relative to the center and the other end moving away. The rotational velocities are measured along the galactic diameter represented by our view of the disk by observing the variations in redshift, those variations being a Doppler effect. Galactic rotation curves so obtained do not exhibit the expected Keplerian form, an inverse square root of radius. Rather, they exhibit rotational velocity independent of radius. The overall curve, after the portion pertaining to the dense central core of the galaxy, is a transition to a flat curve in the region corresponding to the spread-out galactic disk as in Figure D-4, below.

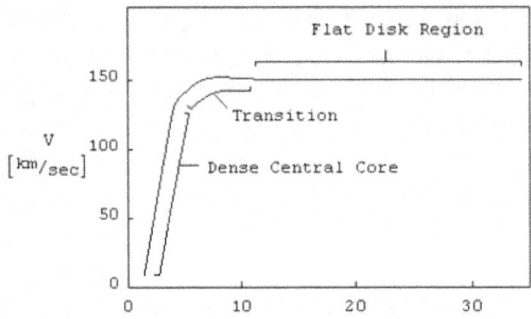

Figure D-4 - A Typical Galactic Rotation Curve as Observed

Because the form of the flat portion of galactic rotation curves lies between the case of a dominant central mass, as the Keplerian inverse square root of radius form [Figure D-1], and the case of a uniformly dense mass, with its direct proportion to radius form [Figure D-2], it has been inferred that matter that we have not observed must be present similarly distributed within the galaxy. That is, it is inferred that unobservable matter must be distributed in the galaxy in a manner that lies between the matter distribution of a dominant central mass [the Keplerian case] and that of a uniformly dense mass [the direct proportion to radius case] as a halo of "dark matter" which causes the rotation form that exhibited. Thus the "dark matter" hypothesis.

No explanation has been offered for why the "dark matter", while performing a gravitational function in the galaxy nevertheless fails to be distributed in the same manner as the "visible matter", a more extensive flat disk which has a far smaller density of more widely dispersed stars.

However, what the rotation curves demonstrate is not the existence of a hypothesized cause [dark matter]; <u>they only demonstrate the existence of an acceleration that is not accounted for</u>. That acceleration is identified as follows. A constant acceleration, $\Delta a_{Anomalous} = 8.7 \cdot 10^{-8}$ cm/sec² = a_A, acting alone as a gravitational acceleration maintaining a mass in orbit, would produce a rotation curve as in Figure D-5, below.

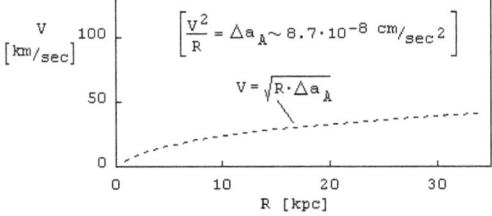

Figure D-5 - The Rotation Curve of $a_{Anomalous}$ Acting Alone

That rotation curve is of the correct form and magnitude to convert a galactic rotation curve of Keplerian form [as in Figure D-1] to a flat one [as in Figure D-4]. That is, the rotation curve of $a_{Anomalous}$ exhibits V <u>directly</u> proportional to the square root of R and the Keplerian rotation curve exhibits V <u>inversely</u> proportional to the square root of R. The two effects tend to cancel and leave a flat rotation curve. With the naturally occurring typical rotation curve modified by the addition of $a_{Anomalous}$ the rotation curve becomes flat, as illustrated in Figure D-6, below, by superimposing the curves.

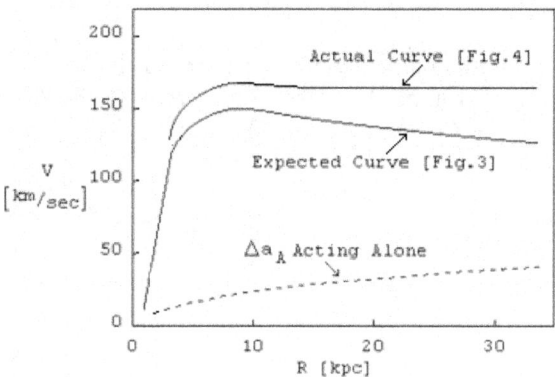

Figure D-6 - The Anomalous Acceleration, $a_{Anomalous}$, Acting Alone Superimposed on the Expected and Actual Rotation Curves [Figures D-3 & D-4]

Of course, the rotational velocities corresponding to the components of the total acceleration cannot properly be added. Rather, the accelerations must be summed and the resulting rotational velocities then obtained as follows,

(D-5) Total Acceleration = "natural acceleration" + Δa_A

$$\frac{v^2}{R} = \frac{G \cdot M}{R^2} + \Delta a_A \quad \text{so that} \quad v = \left[\frac{G \cdot M}{R} + R \cdot \Delta a_A \right]^{\frac{1}{2}}$$

which produces the observed actual flat portion of the rotation curve in the region corresponding to where the "expected" form is Keplerian.

The Pioneer Anomaly

The preceding galactic rotation curve anomalous acceleration $\Delta a_{Anomalous} = 8.7 \cdot 10^{-8}$ cm/sec2 is identical in magnitude to the Pioneer Anomaly anomalous acceleration. The Pioneer Anomaly is a small acceleration of $8.7 \cdot 10^{-8}$ cm/sec2, centrally directed [toward the Sun], constant, distance independent, and of unknown cause. The evidence for it is abundant tracking data that have been reviewed and re-reviewed in search of error with the result that the effect is highly validated.

Since the original reporting of the Pioneer Anomaly in 1998 sources of systematic error external to the spacecraft [e.g. solar wind / radiation], internal to the spacecraft [e.g. gas leakage], and in the computational system [e.g. model accuracy / consistency] have all been thoroughly examined. All of those sources of error are either too small, not applicable, and/or act in the wrong direction to account for the phenomenon. The input of suggested sources of systematic error to those analyses has been not only from the research team of authors but from a number of other sources interested in the problem. The source area of systematics has been essentially exhausted.

The only difference between the Pioneer Anomaly acceleration and the galactic rotation curve anomalous acceleration is that in the Pioneer case the acceleration is directed toward the Sun, the dominant factor in the mechanics of the Pioneer spacecrafts' motion whereas the galactic rotation curve anomalous acceleration is directed toward the rotational center of the galaxy, the dominant factor in the mechanics of galaxy rotation.

The Flybys Anomaly

In March 2008 anomalous behavior in spacecraft flybys of Earth was reported in Physical Review Letters, Volume 100, Issue 9, March 7, 2008, in an article entitled "Anomalous Orbital-Energy Changes Observed during Spacecraft Flybys of Earth"[1].

The data indicate unaccounted for changes in spacecraft speed, both increases and decreases, for six different spacecraft involved in Earth flybys from December 8, 1990 to August 2, 2005. These anomalous energy changes are a function of the incoming and outgoing geocentric latitudes of the asymptotic spacecraft velocity vectors and further indicate that a latitude symmetric flyby does not exhibit the anomalous speed change. The article states that, "All ... potential sources of systematic error .... [have been] modeled. None can account for the observed anomalies.... "Like the Pioneer anomaly ... the Earth flybys anomaly is a real effect .... Its source is unknown."

A phenomenon like that involved in galactic rotation curves and in the Pioneer Anomaly would account for the highly varied occurrences of the flyby anomaly: a small acceleration [in addition to that of natural gravitation], centrally directed and independent of distance; that is a modest and otherwise unknown acceleration directed toward the core center of the Earth, the principle body involved, the dominant factor in the mechanics of the flyby.

To observe the relation to the Flybys Anomaly of an otherwise unknown or undetected anomalous, centrally directed, distance independent acceleration the first step is to consider a simple spacecraft pass of Earth where the pass is all at zero latitude as shown in Figure D-7, on the following page. In the vectors analysis part of the figures A is the full anomalous acceleration, C is its component parallel to the direction of motion of the satellite, and θ is the angle between the direction of action of those two.

When the spacecraft is at a great distance out from Earth the spacecraft's motion is close to being directed toward the center of the Earth but not exactly so. A centrally directed acceleration there analyzed into components parallel and perpendicular to the spacecraft's motion would show most of the centrally directed acceleration acting to increase the spacecraft's speed.

As the spacecraft travels nearer to Earth that component parallel to its motion decreases, becoming zero at the closest approach to Earth. From that point on the parallel component acts in the opposite direction on the spacecraft, that is its effect is to decelerate the spacecraft not accelerate it. Ultimately the anomalous acceleration and anomalous deceleration experienced by the spacecraft become equal and cancel each other out leaving as the only flyby effect the gravitational boost, due to another effect, that is the overall purpose of the flyby.

Of the full centrally directed acceleration, A, the component, C, parallel to the path of the flyby in this case is

$(D-1) \quad C = A \cdot \cos[\theta]$

which is apparent if the flyby path is a straight line. However, the actual flyby path is somewhat curved by the Earth's gravitation. But, the anomalous acceleration is always centrally directed toward the core of the Earth so that C is nevertheless as stated.]

a. Polar View - Flyby

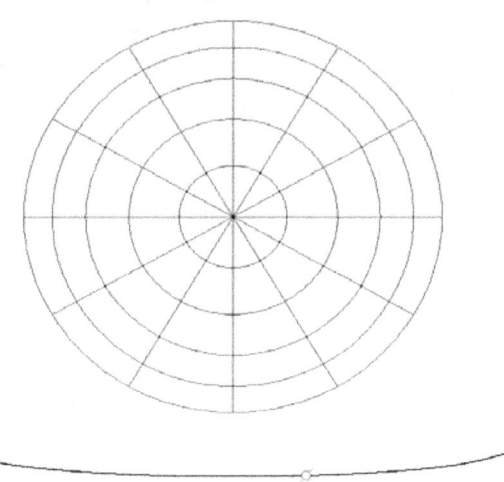

b. Polar View - Anomalous Acceleration Vectors

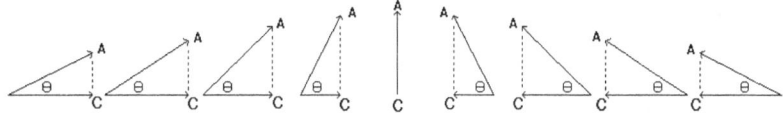

Here the acceleration phase and the deceleration phase are equal and offset each other.

c. Equatorial View

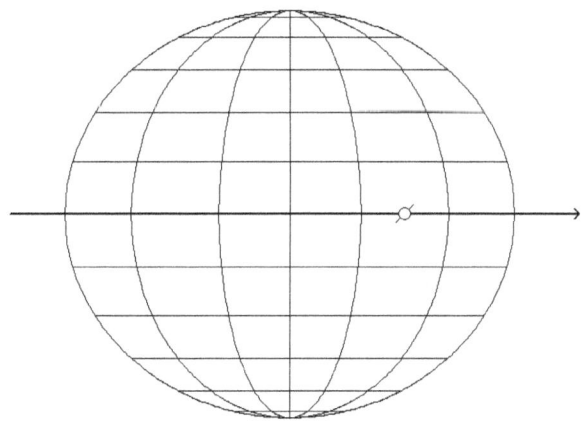

Figure D-7
A Zero Latitude Pass

Equation *D-1* is valid when the flyby pass is solely at zero latitude. However, if other than zero the latitude of the flyby pass has a significant effect on the magnitude of C, the component of the overall centrally directed acceleration parallel to the spacecraft flight path. As latitude increases the magnitude of C, decreases. That is most easily visualized by imagining the flyby over the geographic north pole at 90° north latitude. There the centrally directed acceleration toward the center of the Earth has no component parallel to the flight path.

Therefore, for flyby paths at other than zero latitude the effective value of A is $A(\lambda)$ a function of latitude, λ, as Equation D-2

(D-2)   A = A(λ) = A·Cos[λ]

so that Equation D-1 then becomes Equation D-3 the full expression for the extent to which the centrally directed anomalous acceleration actually accelerates or decelerates the spacecraft.

(D-3)   C = A·Cos[λ]·Cos[θ]

The gross effect of latitude can be evaluated by examining three cases:

> A - The flyby path is symmetrical relative to the equator so that the latitude effect in the first half of the flyby, θ = 0° to 90°, is exactly offset or balanced by the second half of the flyby, θ = 90° to 180°. This case is essentially the same as presented in Figure D-7, above.
>
> B - The flyby path starts at low latitude and finishes at high latitude, Figure D-8 on the following page.
>
> C - The flyby path starts at high latitude and finishes at low latitude, Figure D-9 on the second following page.

Per the equations and Figure D-7 in the first half of the flight path the effect of the anomalous, centrally directed acceleration is to increase the speed of the spacecraft whereas the effect in the second half of the flight path is to decrease the spacecraft's speed. By its definition Case A produces no net anomalous acceleration or deceleration of the spacecraft because the first and second halves of the flight path balance and offset each other.

In Case B, the first half, i.e. the acceleration half, of the flight path is at low latitude where the latitude effect only modestly reduces the anomalous acceleration magnitude. But for that case and path the second half, i.e. the deceleration half, of the flight path is at a high latitude where the latitude effect greatly reduces the anomalous acceleration magnitude. The net effect is a relatively large acceleration followed by a lesser deceleration for a net increase in the spacecraft's speed.

In Case C, the effect is just the reverse of that in Case B; the first, i.e. the acceleration, half of the flight path is at high latitude where the effect of the latitude greatly reduces the anomalous acceleration magnitude. But for that case and path the second, i.e. the deceleration, half of the flight path is at a low latitude where the effect of the latitude only modestly reduces the anomalous acceleration magnitude. The net effect is a relatively small acceleration followed by a greater deceleration for a net decrease in the spacecraft's speed.

Therefore, depending on the specific flight path of the spacecraft's flyby pass of Earth the spacecraft may experience an overall net anomalous acceleration or a net anomalous deceleration, those in various amounts depending on the specific encounter and the latitudes involved, and zero net modification if the path is perfectly latitude symmetrical.

D – THE UNIVERSAL EXPONENTIAL DECAY

a. Equatorial View - Flyby

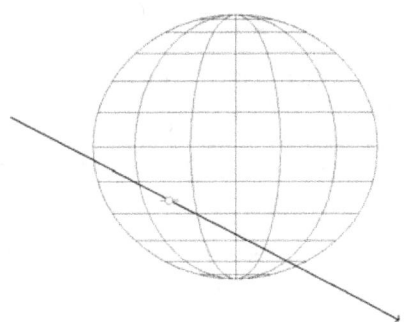

b. Equatorial View - Flyby, Rotated

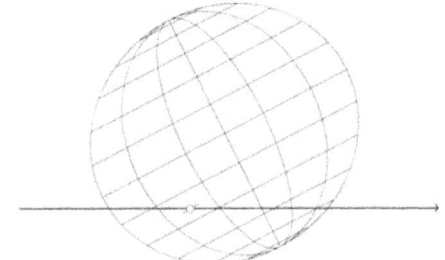

c. Anomalous Acceleration Vectors

Vectors As In zero Latitude Pass Case

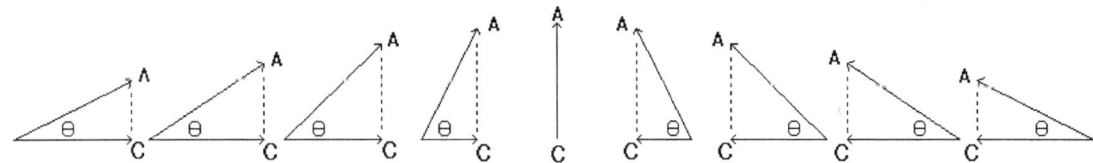

Above Vectors as Further Reduced by Non-Zero Latitude
Reduction Factor = Cosine[Latitude]

Latitude:								
10°	20°	30°	40°	50°	60°	70°	80°	90°
Factor:								
.99	.94	.87	.77	.64	.50	.34	.17	.00

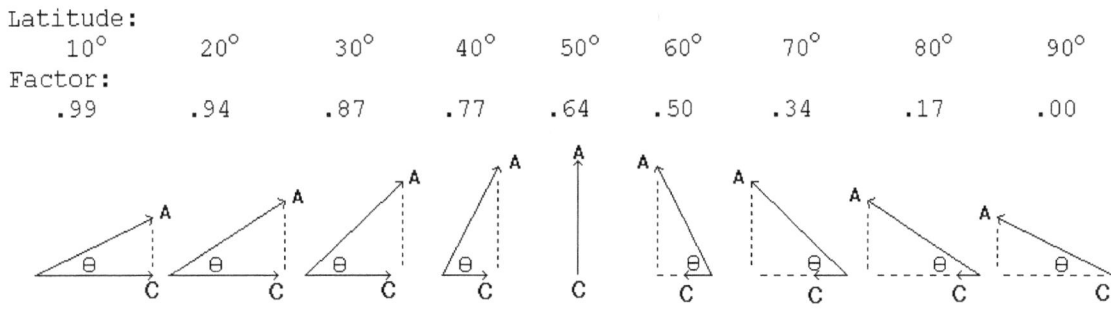

```
The result in this case is a net acceleration
[to the right in the diagrams].
```

Figure D-8
A Pass at Increasing Latitude

a. <u>Equatorial View - Flyby</u>

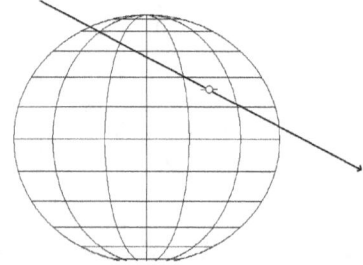

b. <u>Equatorial View - Flyby, Rotated</u>

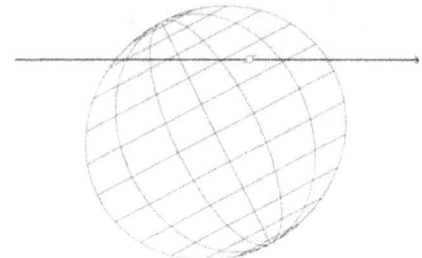

c. <u>Anomalous Acceleration Vectors</u>

Vectors As In zero Latitude Pass Case

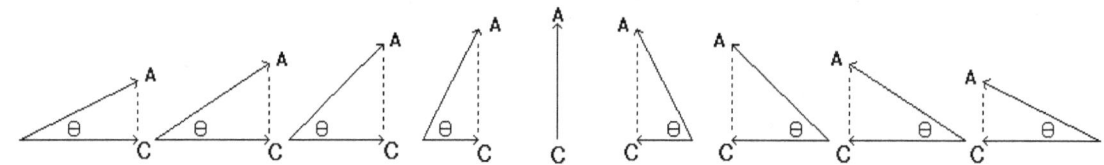

Above Vectors as Further Reduced by Non-Zero Latitude
Reduction Factor = Cosine[Latitude]

Latitude:								
90°	80°	70°	60°	50°	40°	30°	20°	10°
Factor:								
.00	.17	.34	.50	.64	.77	.87	.94	.99

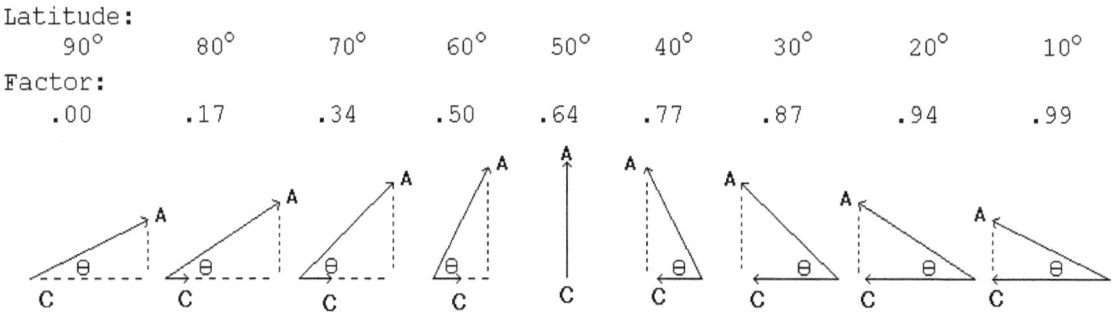

The result in this case is a net deceleration, that is an acceleration toward the left in the diagrams, against the direction of the flyby.

Figure D-9
A Pass at Decreasing Latitude

The "Dark Flow" Anomaly

Thus there are small, centrally directed, distance independent, non-gravitational, same, anomalous accelerations appearing as a near Earth effect [the Flybys Anomaly], a Solar effect [the Pioneer Anomaly], and a galactic effect [galactic rotation curves]. It can only be concluded that the same effect must appear relative to every planet [and every planet's moons], every sun [star], every galaxy and every group of galaxies.

And such a small, centrally directed, distance independent, non-gravitational, same, anomalous acceleration could be expected to appear for the universe overall, directed toward the center of the universe, the location of the origin, where it all began.

The universe began with the "Big Bang", an immense explosion radially outward in all directions, largely spherically symmetrically, from an original source "singularity".

We, residing on planet Earth, of star Sol, in one of several branches of spiral galaxy Milky Way, are located off some significant distance in "our general direction" from and relative to the location of the original singularity.

We can "see" or detect a large number of neighbor galaxies, distant and near, whose components similarly proceeded outward from that "Big Bang" in directions slightly or significantly other than our particular direction.

But, there is a further mass of stellar bodies that proceeded outward from the "Big Bang" in directions away from us. What we can detect is only well less than half the total product of the "Big Bang".

The original location of the singularity, the origin, lies essentially at the center of the largely spherical volume of the source's product, the expanding universe. And the universe that we "see" lies largely to one side of that origin's location

To the above list of three effects caused by the systematic contraction of the universe can now be added the "Dark Flow" as originally reported in 2008 and recently further analyzed in terms of extensive new data as reported in NASA Goddard Release No.: 10-023 and in A. Kashlinsky, F. Atrio-Barandela, H. Ebeling, A. Edge, and D. Kocevski. *A New Measurement of the Bulk Flow of X-Ray Luminous Clusters of Galaxies.* [7].

Distant galaxy clusters mysteriously stream at a million miles per hour towards a single point in the sky, separate from the expansion of the universe, along a path roughly centered on the southern constellations Centaurus and Hydra. A new study led by Alexander Kashlinsky at NASA's Goddard Space Flight Center in Greenbelt, Md., tracks this collective motion -- dubbed the "dark flow"....

The clusters appear to be moving along a line extending from our solar system toward Centaurus / Hydra ... away from Earth. The distribution of matter in the observed universe cannot account for it. Its existence suggests that some structure beyond the visible universe -- outside our "horizon" -- is pulling on matter in our vicinity.

This is indication of the overall universe's experiencing an anomalous centrally directed acceleration accelerating all the matter of the universe gradually back toward the location of its origin [as described above]. This "Dark Flow" is part of that centrally directed acceleration toward the location of the origin of the universe, a location at or just beyond the "edge" of the universe "see-able" by us.

A "map" of the universe that we "see" would look somewhat as in Figure D-10, below, where the regions of galaxies studied involved in the "dark flow" are indicated in the large colored areas in red, yellow, green and blue.

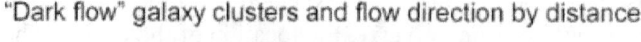

"Dark flow" galaxy clusters and flow direction by distance

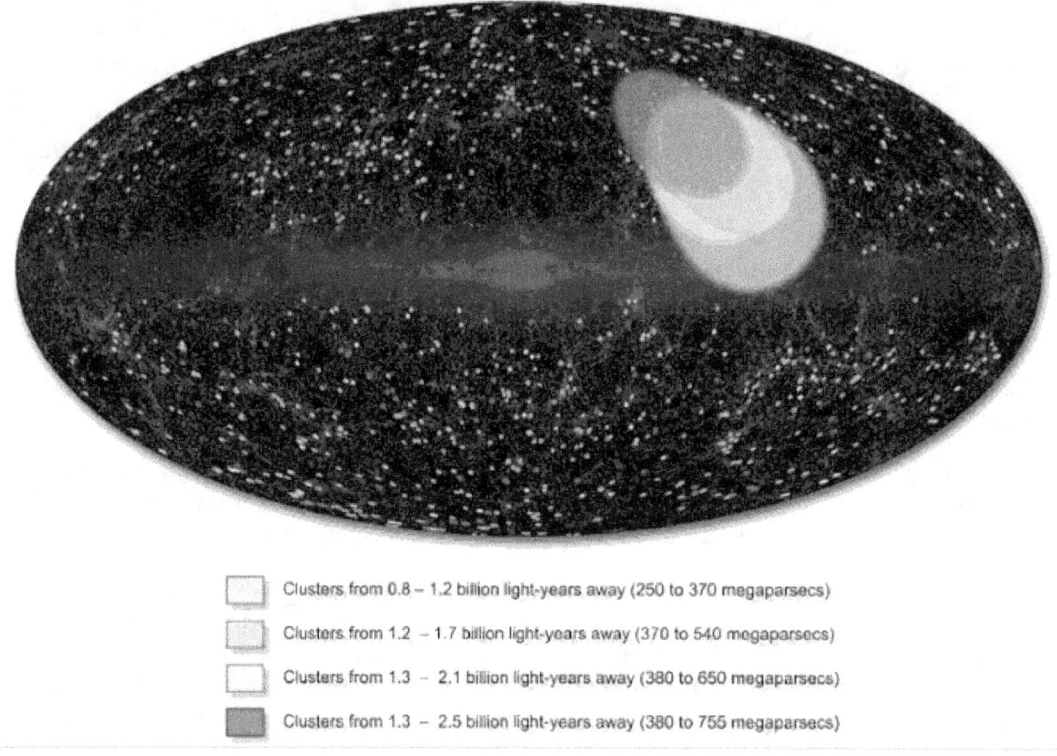

The coloured dots are clusters within one of four distance ranges, with redder colours indicating greater distance. Coloured ellipses show the direction of bulk motion for the clusters of the corresponding colour. (Courtesy: NASA/Goddard/A.Kashlinsky et al.)

Figure D-10
A Map of That Part of the Universe Observable to Us

A General Anomalous Acceleration Throughout the Universe

Thus there are small, centrally directed, distance independent, non-gravitational, same, anomalous accelerations appearing as a near Earth effect [the Flybys Anomaly], a Solar effect [the Pioneer Anomaly], a galactic effect [galactic rotation curves], and a Universe effect [the Dark Flow]. It can only be concluded that the same effect must appear relative to every planet [and every planet's moons], every sun [star], every galaxy and group of galaxies, and the universe overall. In other words as a general cosmic effect.

What could produce such a phenomenon ? What would cause there to be a universe-wide occurrence of such same accelerations ?

Taken together, planet relative, star relative, galaxy relative, universe relative, they collectively are a systematic contraction, a gradual reduction in the length component of every physical quantity in the universe. A general universal decay.

In material reality such decays are exponential. There are myriad examples of such, for example: radioactive decay, the decay of electrical transients in circuits involving inductance and capacitance, the decay of motion transients in mechanical systems involving mass and spring, the amplitude decay in a rung bell or a plucked string, etc. It is not unreasonable that a universe that began with an explosive "bang" follow that with a gradual exponential decay.

Such a decay of the overall universe was predicted and analyzed in detail in 1998, before the appearance of all except the earliest of the foregoing various anomalies, in the book *The Origin and Its Meaning*, Section 21 [5].

The Universal Exponential Decay is an exponential decay of the length dimensional aspect of all quantities in the universe. It involves the fundamental constants (c, q, G, h, etc.) and decay of any of those must be dimensionally consistent with the decay of the others. The dimension that is decaying is length, the $[L]$ dimension in the dimensions of, for example: h, $[M \cdot L^2/T]$; c, $[L/T]$; and G, $[L^3/M \cdot T^2]$. The time constant of the decay is about $\tau = 3.57532 \cdot 10^{17}$ sec ($\approx 11.3373 \cdot 10^9$ years).

Objections that such an effect would conflict with the known planetary system performance per the highly accurate planetary ephemeris are a mistaken interpretation of the situation. Consider a planet in circular orbit around a sun as in Figure D-11, below.

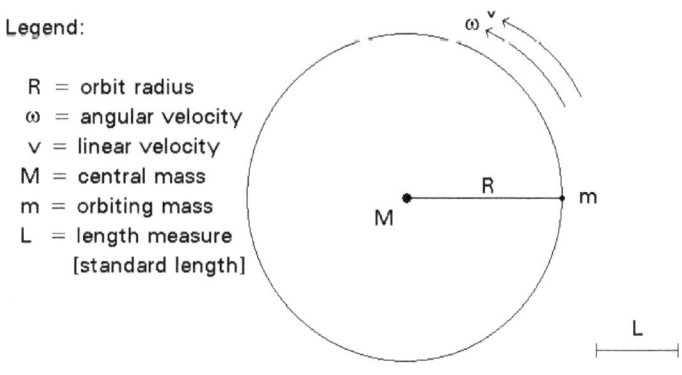

Legend:

R = orbit radius
ω = angular velocity
v = linear velocity
M = central mass
m = orbiting mass
L = length measure
   [standard length]

Figure D-11

The relationship governing the motion is, of course, Equation D-4, below

(D-4) Centripetal Acceleration Required = Gravitational Attraction Acceleration

$$v^2/R \; (or) \; R \cdot \omega^2 \quad = \quad G \cdot M/R^2$$

Now, let the length dimensional aspect [with the dimensions of all quantities expressed in the fundamental dimensions of mechanics, $[L]$, $[M]$, and $[T]$] of all quantities decay, becoming gradually smaller with time. That is, let all lengths, [L], decrease by being multiplied by the decay function, D(t), per Equation D-3, below. [For the present purpose the form of the decay function is irrelevant except that it must be a function of time. The decaying exponential is used because it is common in nature and is a complicated case.]

(D-5) $D(t) \equiv \varepsilon^{-[t/\tau]}$, where τ is the time constant of the decay

Then the quantities involved in Equation D-4 all change to as follows.

(D-6) <u>The Orbital Radius, R, [dimension = **L**]</u>

R becomes $R(t) = R(t=0) \cdot \varepsilon^{-[t/\tau]}$

<u>The Gravitational Constant [dimensions = $L^3/_{M \cdot T^2}$]</u>

G becomes $G(t) = G(t=0) \cdot \{\varepsilon^{-[t/\tau]}\}^3$

<u>The Centripetal Acceleration Required [dimensions = $L/_{T^2}$]</u>

$R \cdot \omega^2$ becomes $R(t) \cdot \omega^2 = [R(t=0) \cdot \varepsilon^{-[t/\tau]}] \cdot \omega^2$

$$= [R(t=0) \cdot \omega^2] \cdot \varepsilon^{-[t/\tau]}$$

or

$$\frac{v^2}{R} \text{ becomes } \frac{[V(t)]^2}{R(t)} = \frac{[V(t=0) \cdot \varepsilon^{-[t/\tau]}]^2}{[R(t=0) \cdot \varepsilon^{-[t/\tau]}]}$$

$$= \frac{[V(t=0)]^2}{R(t=0)} \cdot \varepsilon^{-[t/\tau]}$$

<u>The Gravitational Attraction Acceleration</u> [dimensions = $L/_{T^2}$]

[and where the G dimensions = $L^3/_{M \cdot T^2}$]

$$\frac{G \cdot M}{R^2} \text{ becomes } \frac{G(t) \cdot M}{[R(t)]^2} = \frac{[G(t=0) \cdot \{\varepsilon^{-[t/\tau]}\}^3] \cdot M}{[R(t=0) \cdot \varepsilon^{-[t/\tau]}]^2}$$

$$= \frac{G(t=0) \cdot M}{[R(t=0)]^2} \cdot \varepsilon^{-[t/\tau]}$$

The overall net effect is: R decreases, the required centripetal acceleration decreases in proportion, the gravitational attraction likewise decreases in proportion, and ω is unchanged.

Furthermore, we observers, using our measuring standard ruler, length L of the above Figure D-11, would never detect any of the decay because our standard length would also be decaying at exactly the same rate, in the same proportion.

The point of this obvious mathematics / physics exercise is that a universal decay of the length aspect of all material reality would not conflict with the planetary ephemeris and would not even be detectable at all except in unusual circumstances such as the

Pioneer and Flyby anomalies and the evidence of galactic rotation curves; nor would it interfere with the relative values of the fundamental constants and their interactions in physical laws.

Returning to the orbiting body of Figure D-11, reproduced as Figure D-12 below, the figure's annotations slightly modified, the development of the anomalous acceleration is very direct.

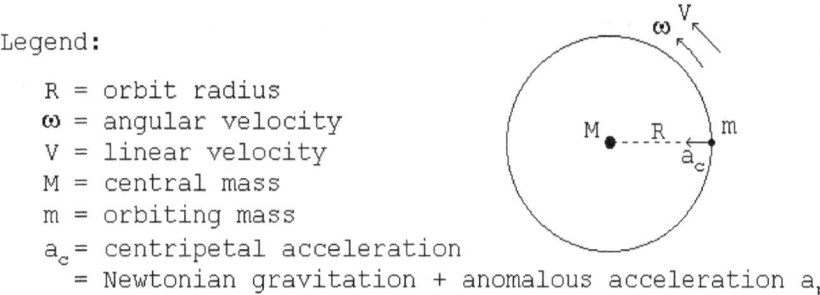

```
Legend:
    R = orbit radius
    ω = angular velocity
    V = linear velocity
    M = central mass
    m = orbiting mass
    a_c = centripetal acceleration
        = Newtonian gravitation + anomalous acceleration a_p
```

Figure D-12

The Newtonian component of the centripetal acceleration is only sufficient to maintain the orbit, to keep R constant, to prevent its increasing. For the orbiting body, m, to gradually approach the central mass, M, that is for R to decrease, additional inward acceleration is required.

That inward acceleration is the anomalous acceleration appearing as a near Earth effect [the Flybys Anomaly], a Solar effect [the Pioneer Anomaly], and a galactic effect [galactic rotation curves]. It is an unavoidable concomitant effect of the contraction of the length dimension *[L]* of R in the above example and of the systematic contraction, the gradual reduction in the length component, of every physical quantity in the universe, of all material reality.

Recent Updates

The Flyby Anomaly has only been detected with a high level of precision in flybys of Earth, due to the availability of monitoring stations such as NASA's in Robledo de Chabela (Madrid) or that of the European Space Agency in Cebreros (Ávila), which allow for the variations in spacecraft speed to be precisely recorded by radar. The latest such Flyby Anomaly instance was the October 9, 2013 Earth flyby by the spacecraft Juno. Flybys of other bodies besides Earth take place but the speed monitoring facilities necessary to detect the anomaly are not available for distant-from-Earth flybys of Saturn and its moon Titan in 2004-2005 by the spacecraft Cassini and of Mars in 2007 by the spacecraft Rosetta.

Summary Conclusions

1 – All four effects: the Galactic Rotation Curves Anomaly, the Pioneer Anomaly, the Flybys Anomaly, and the Dark Flow involve the same common action, a small, centrally directed, non-gravitational, distance independent acceleration, apparently the same common acceleration $\Delta a_{Anomalous} = 8.7 \cdot 10^{-8}\ cm/sec^2$.

2 – The occurrence of such an acceleration apparently universe-wide is indicative of an on-going general contraction of the length aspect of all material reality including the length dimensional aspect of all fundamental constants.

Details on the universal contraction, or decay -- its cause, origin and characteristics are too lengthy for this report and are provided in full in reference [5].

The Universal Decay

The Universal Decay causes the speed of light now to be a smaller, decayed value relative to light speed earlier. Thus in general the speed of light is $c(t) = c_0 \cdot \varepsilon^{-t/\tau}$. [$c_0$ is the original speed of light at the instant of the "Big Bang" and t is time since the "Big Bang"].

The speed of light is now decaying from its present value as we know it, c or c_{now}, as $c(t) = c_{now} \cdot \varepsilon^{-t/\tau}$. Therefore the rate of change of the speed of light now is as follows.

$$(D\text{-}7) \quad \frac{d[c(t)]}{dt} = -\frac{c_{now}}{\tau} = -\frac{2.99792 \cdot 10^{10}}{3.57532 \cdot 10^{17}} = -8.38504 \cdot 10^{-8} \text{ cm}/s^2$$

compared to the Pioneer Anomaly $= -(8.7 \pm 1.33) \cdot 10^{-8} \text{ cm}/s^2$

That rate of change of the speed of light is due to the rate of change of its length dimensional aspect and, therefore, is the at present rate of change of all length dimensional aspects. It is the rate of the universal contraction, the un-accounted for centrally directed acceleration demonstrated in galactic rotation curves, the Flybys Anomaly and the Dark Flow Anomaly.

Because the decay time constant is so large the at-present rate appears to us to be constant.

Because everything including our instrumentation, our measurement standards, our atoms and ourselves are all experiencing the same decay, the decay is unnoticeable to us and is generally undetectable by us except for unusual circumstances such as the anomalies presented above.

Validating The Universal Decay

Because the speed of light is decaying, light emitted long ago is faster than our present contemporary light, which causes the ancient light to appear to us to have a longer wavelength, that is, to be Redshifted. [Some of Redshifts, but not more than a minor portion, is due to the Doppler Effect of the astral sources' outward velocities.]

Aside from observation of Redshifts, <u>each observation of which is actually an observation of the universal decay</u>, there are two other specific experimental observations that can be conducted to verify the Universal Decay and the value of its decay time constant.

- It can be tested that the speed of the light from distant astral sources is larger than our contemporary light speed. The earlier procedure of Michaelson or Pease and Pearson using the Foucault method is now superseded by the modern procedure, which is to modulate the light beam and use that modulation to measure the time required for the light to traverse a known distance.

- It can be tested that the Planck Constant of the light from distant astral sources is larger than our contemporary Planck Constant, h, using the photoelectric effect. Measuring the retarding potential that reduces the photoelectric current to zero, for light spectrally selected of a specific frequency, plots [for a set of different frequencies] as diagonal straight lines whose slope is the Planck Constant of that light.

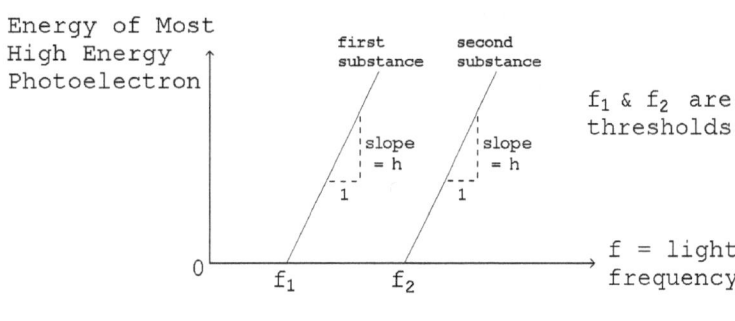

Figure D-13

References

[1] Zwicky, F. (1933), *Die Rotverschiebung von extragalaktischen Nebeln*, Helvetica Physica Acta **6**: 110–127, may be found at http://adsabs.harvard.edu/cgi-bin/nph-bib_query?bibcode=1933AcHPh...6..110Z.

[2] J. D. Anderson, P. A. Laing, E. L. Lau, A. S. Liu, M. M. Nieto, and S. G. Turyshev, *Indication, from Pioneer 10/11, Galileo, and Ulysses Data, of an Apparent Anomalous, Weak, Long-Range Acceleration*, Phys. Rev. Lett. **81**, 2858 (1998).

[3] John D. Anderson, James K. Campbell, John E. Ekelund, Jordan Ellis, and James F.Jordan, *Anomalous Orbital-Energy Changes Observed during Spacecraft Flybys of Earth*, Physical Review Letters, PRL **100**, 091102 (2008).

[4] A. Kashlinsky, F. Atrio-Barandela, D. Kocevski, H. Ebeling, *A Measurement of Large-Scale Peculiar Velocities of Clusters of Galaxies: Results and Cosmological Implications*, Astrophysical Journal Letters, Print edition October 20, 2008; online week of September 22, 2008.

[5] R. Ellman, *The Origin and Its Meaning*, The-Origin Foundation, Inc., http://www.The-Origin.org, 1996. [The book may be downloaded in .pdf files from http://www.The-Origin.org/download.htm].

[6] R. Ellman, *Analysis of the "Big Bang" and the Resulting Outward Cosmic Expansion: Hubble - Einstein Cosmology vs. The Universal Exponential Decay*, available at http://www.arXiv.org, arXiv:physics/0004053 [pdf].

[7] A. Kashlinsky, F. Atrio-Barandela, H. Ebeling, A. Edge, and D. Kocevski. *A New Measurement of the Bulk Flow of X-Ray Luminous Clusters of Galaxies*. The Astrophysical Journal, 2010; 712 (1): L81 DOI: 10.1088/2041-8205/712/1/L81

Appendix E

The 'Big Bang' Outward Cosmic Expansion

PART I -- THE THEORIES

1 - The Nature of Space

There is general agreement that the universe began with and results from an essentially instantaneous appearance and outward "explosion" of the matter and energy of the universe at a "singularity", and the resulting on-going expansion -- the "Big Bang".

a. Hubble - Einstein Theory

The Hubble - Einstein theory, thoroughly and extensively elaborated in numerous books and scientific papers, is that the result of that beginning was the creation of space, itself, and that it is space, itself, that is expanding, and in the process carrying the universe's matter and energy along with it -- an expanding universe such that the velocity, v, of recession of any distant astral object from us, the observers, is directly proportional to the object's distance from us, $v = H_0 \cdot d$, where H_0 is the "Hubble Constant", the value of which has been not well determined beyond being in the range of 65 - 80 km/sec *per megaparsec*, but has been recently reported per analysis of a Hubble Space Telescope survey as 72 km/sec *per megaparsec*.

In spite of the long term acceptance of the Hubble - Einstein cosmological concept there are fundamental questions about it that are unanswered. The concept is a direct result of Einstein's General Theory of Relativity for which space, itself, is some kind of "substance" [not Einstein's terminology] capable of expanding and capable of being "curved" by the effect on it of gravitating masses in it. That concept leaves the problem, "... relative to what" ? If space is expanding then the expansion must be relative to some static, non-expanding reference. If space is curved than the curvature must be relative to some flat, uncurved reference. One cannot have relativity without relativity. Any change or effect must be relative to a previous unchanged reference or previous unaffected state. Otherwise the change or effect would be undetectable.

So, what do we call that "static, flat, uncurved reference"? It is space itself; and it is, and it must be, the framework that expansion of the universe is relative to. And flat, uncurved rectilinear space is and must be, the framework that curved motion due to gravitation is relative to. And that space must have always existed unoccupied [and, therefore actually "nothing"] until the "Big Bang" introduced matter and energy into it.

Furthermore, were space, itself, to be expanding as in the Hubble - Einstein theory, then it would be expanding everywhere including the expansion of the space containing and within all of our measurement standards and instrumentation [and selves]. But, an expanding ruler used to measure an expanding universe would report only a static state, not an expansion. The expansion would not be detectable by us if it were space, itself, that is expanding. Since we detect the expansion, then it must be, and is, the objects within space that are moving away from each other [away from the "Big Bang" location]. And, therefore, space itself is passive and static.

b. The Universal Decay Theory

The Universal Decay theory that the universe is exponentially decaying while its content is expanding outward within static "space" was first propounded in "*The Origin and Its Meaning*" in 1996 and has since been validated by the Pioneer 10 and 11 "anomalous acceleration" as well as by the theory's success in accounting for "dark matter" and "dark energy". As in Appendix D the same centrally- directed, distance-independent acceleration of $(8.74 \pm 1.33) \times 10^{-8}$ cm/s^2 that is the Pioneer "anomalous acceleration" supplies the "additional gravitation" that "dark matter" is sought to supply and is part of the on-going contraction and decay of all length, $[L]$, dimensions. The details, mechanisms, and parameters are all thoroughly developed in the references. For the present purposes the pertinent aspects are as follows.

The Universal Decay theory is that the length, $[L]$, dimensional aspect [with all dimensions expressed in the fundamental dimensions of mechanics, $[L]$, $[M]$, and $[T]$] of all quantities in the universe [e.g. distance $[L]$, speed $[L/T]$, gravitation constant, G, $[L^3/M \cdot T^2]$, Planck constant, h, $[M \cdot L^2/T]$, etc.] is exponentially decaying with time in the form:

$$(E\text{-}1) \quad L(t) = L(0) \cdot E^{-t/\tau}$$

and the value of the time constant, τ, [predicted in "*The Origin and Its Meaning*" from theoretical calculations and validated by the Pioneer 10 and 11 "anomalous acceleration"] is

$$(E\text{-}2) \quad \tau = 3.57532 \cdot 10^{17} \text{ sec} \quad [\approx 11.3373 \cdot 10^9 \text{ years}]$$

Because it is the length, $[L]$, dimensional aspect of all quantities that is decaying, all of those quantities remain consistent with each other through the laws of physics that connect them. The very long time constant compared to a human life time would make our observation of the decay difficult. And, in any case, because all of our measuring instrumentation and we ourselves are undergoing that same decay, we cannot directly measure or observe it.

One would expect that exactly the corresponding argument to the above presented last criticism of the hypothesis that space, itself, is expanding -- specifically that the Universal Decay should be undetectable because the measuring equipment is also decaying -- would prevent ever observing the decay. However, in the case of the Pioneer 10 and 11 satellites and the case of galactic rotation curves, the decay has been detected because it forces orbital / path behavior that would not be present if there were no decay, and that orbital / path behavior can be and has been observed.

The Universal Decay is the principal cause of redshifts. There must be some Doppler content in redshifts because the astral sources do have velocities away from us, the observers, but it can readily be shown that the Doppler effect accounts for only about 10% or less of the total redshift. The Universal Decay produces the redshifts because when we observe light from distant astral sources we are observing light emitted long ago, which means that it was emitted less decayed than our local contemporary light. Less decay of the length, $[L]$, dimensional aspect of all quantities in that light means that its speed and wavelength are greater than the spectrally corresponding light from sources local to us. Being less decayed it appears redshifted to us because we compare it to our local, decayed-to-date light and spectra. The decay is of the length, $[L]$, dimensional aspect, which affects the wavelength. There is no decay of the period, $[T]$, nor its inverse, the frequency.

There is no violation of invariance involved. At any instant of time the speed of the light emitted by any and every source anywhere and everywhere in the universe is the same, that is the as-of-that-time current decayed value, because all decay started at the same time, the instant of the "Big Bang", and all decay is with the same decay constant, τ. However, once emitted the light continues propagating onward at the speed at which it was emitted. The decay is in the generation, is in the source, not the propagation. That is the case because the emitted light carries within it its own propagation-determining permeability and dielectric constant, μ and ε. How could it be otherwise since light propagating outward into unoccupied space, into pure nothing, would certainly find no μ and ε there: in nothing?

As a consequence, therefore, direct measurement of the speed and Planck's constant $[c$ and $h]$ in ancient light is another way to observe the decay because it applies our decayed equipment to light that is much less decayed.

c. Comparison of the Two Theories

A quick calculation makes clear that, in a universe in which the redshifts are dominantly due to the Universal Decay with only a minor [Hubble - Einstein] Doppler content, the age of the universe must be considerably greater than the current estimates, that are based on Hubble - Einstein reasoning, of about *13.7 billion years*. The detection of redshifts of $z \approx 10$ has recently been reported. Those would require an age of the universe [a time period to encompass sufficient multiples of the decay time constant τ] of about *30 billion years*.

In Hubble - Einstein reasoning, the recent observations at those increasingly greater redshifts are nearing the point of leaving little or no time, between the instant of the very beginning and the first appearance of the most ancient galaxies, little or no time for the "Dark Ages" and initial galaxy formation to have taken place -- a process estimated only a few years ago to have required and have taken *2 - 3 billion years* and now reduced by default subtraction [the Hubble - Einstein estimated age of the universe less the Hubble - Einstein estimated age of high redshift light observed] to as little as a *few 100 million years*. And, observation of still greater redshifts with the resulting further compression of the [Hubble - Einstein] time available for the "dark ages" may be shortly forthcoming.

Further in favor of the Universal Decay theory, is that exponential decay is found essentially everywhere in physics, in nature. It would almost seem to be a requirement of a universe coming into existence with a sudden "bang". The Hubble - Einstein cosmological concept appears to be in severe and increasing trouble. It has always suffered from the absence of competition with an alternative "Big Bang" cosmology [its only significant competition was the now defunct steady state universe theory]. In addition, the useful Occam's Razor [the simplest explanation is most likely the correct one] is certainly against Hubble - Einstein.

2 - The Topology of the Universe

a. Hubble - Einstein

In the Hubble - Einstein conception the universe that arose from the "Big Bang" has no "center" and no "edge"; rather, space -- the universe -- is a topologically "closed space" with nothing else beyond its "spatial limits". And, therefore, it is finite.

The theory's space having come into being from the original singularity and having continuously expanded thereafter, it is difficult to justify the concept that the location of the original singularity is nowhere, and in particular that it is not somewhere within the expanding space of the universe that arose from it. Likewise, it is difficult to justify the concept of the expanding universe of space having no edge, no boundary. Because it is expanding, at any moment the universe's "closed space" encloses a smaller "closed space" that it was a moment ago. Some distinction between the two is necessary else there would be no expansion. The enclosed smaller "closed space" must have a boundary or "spatial limits" that distinguish it from the total enclosing larger "closed space" with its larger "spatial limits". If the time interval between the two approaches zero the boundaries of the enclosed become the boundaries of the enclosing.

Or, is the evidence of the expansion the increasing separation distances of the objects within the "closed space"? If so there is no difference between that and the objects traveling outward from their original source through a passive, static space that is not a "closed space".

Thus, while a boundaryless "closed space" can be theoretically conceived of, such a space that is non-static, that is expanding, is impossible as a material reality. Doing away with a boundaryless "closed space" universe with nothing outside of it returns the discussion to the problem: If the universe is finite then what is outside of it, and should not that be called the universe, and must not the universe then be infinite ? But a material infinity is impossible -- how are the two reconciled ? See "The Origin and Its Meaning", below.

Theorist mathematicians like to use analogies to justify their space with no edge. They cite the surface of a sphere as a two-dimensional space having no edge, no boundary, as it exists in the three-dimensional space of the sphere whose surface it is, and they then ask that that example be extrapolated to a three-dimensional universe in a four-dimensional space. But the sphere surface is in three dimensions and has two boundaries: [1] the boundary between the surface of the sphere and everything outside of the sphere and [2] the boundary between the surface of the sphere and everything inside the sphere. [Some theorists like to cite the Moebius Strip, an example of quibbling with definitions, not an example of different space.]

b. *The Origin and Its Meaning*

The Origin and Its Meaning conception of the cosmic topology is that space is a three-dimensional Euclidean metric which is nothing until something occupies it. It is now not nothing because part of it is occupied by the matter and energy of the universe. The metric extends infinitely in all directions, but only a finite portion is occupied by the universe. The unoccupied portion is nothing, only a metric.

The exponential decay of the universe is just that; it is not a decay of the metric in which the universe resides. The decay is relative to the metric.

The universe is finite:

- Finite now in that it has been expanding from a point source outward in all directions at a finite speed for a finite amount of time.

- Finite forever, in that the exponential decay of the upper limit on that finite speed, the speed of light, c, makes it be $c(t)$ as Equation *E-3*,

$$(E\text{-}3) \quad c(t) = c(0) \cdot \varepsilon^{-t/\tau}$$

which produces a finite ultimate result as shown in Equation *E-4*.

(E-4)
$$\int_0^\infty c(t)\,dt = c(0)\cdot \tau$$

3 - The Origin of the "Big Bang"

There is general agreement that the universe had to come into being from a preceding "nothing" at a "singularity".

The reason for the universe's arising from nothing is that the alternative is the prior existence of some not-nothing, something, which then must have its own existence accounted for. The only ultimate "beginning" is truly "nothing".

The reason for requiring an original singularity is that otherwise there would be an infinite rate of change in getting from nothing to something. [Yet, getting an entire universe into existence through the portal of a dimensionless singularity is another concern in the overall problem].

[Incidentally, the entire reasoning above demonstrates that we cannot do without metaphysics and that in a sense metaphysics is superior to physics in that metaphysics can address problems not accessible by physics and metaphysics can lay a basis for broad aspects of physics. Physics itself must depend on reproducible verifiable observations; but the validating of hypotheses based on and resulting from such physics data, especially in the absence of competing alternative hypotheses during the development and validation phase, can be subject to error.]

a. 20th Century Physics

The 20th Century Physics theory for the origin of the "Big Bang" does not really exist in that physics insists that its purview extends only to that which can be observed and measured, and we have not and probably can not observe and measure the origin of the "Big Bang".

However, "unofficial" ideas on the subject rely on thinking that is a result of the combination of quantum mechanics and Heisenberg uncertainty where the latter is considered to be actual, real, probabilistic uncertainty, not Heisenberg's original point that measurement is uncertain because the process of measuring changes that which is measured.

The 20th Century Physics thinking is that the "nothing" from which the universe arose was not quite "nothing" because of quantum variations and uncertainty. Rather, the primal "nothing" is deemed a "quantum foam" with particles continuously flashing into and out of existence. That is nevertheless deemed to be "nothing" because on the average the "quantum foam" is neutral or zero. It is theorized to have produced the universe from a fortuitous flashing into existence.

However, a "quantum foam" even though on the average neutral or zero is not quite the same thing as pure "nothing". The existence of the "quantum foam" still needs to be accounted for. 20th Century Physics' contention in that regard is that the primal "quantum foam" naturally always existed by its nature and the laws of quantum mechanics and uncertainty -- that it was as "nothing" as could possibly ever be. How an entire universe of such particles arose in an instant at a "singularity" and so as to proceed on its expansion, remains a problem.

b. The Origin and Its Meaning

The Origin and Its Meaning conception of the origin of the "Big Bang" is that before the "Big Bang" there was nothing -- absolute nothing with no characteristics nor content -- but that even such a nothing cannot have an infinite duration. Consequently the nothing divided into something and an equal but opposite un-something, thus maintaining conservation while changing sufficiently to interrupt the original nothing's otherwise infinite duration. The universe is the combined result of both of the mutually off-setting "halves" acting jointly.

The event was of extremely low probability, but its probability was not zero -- it was possible. Therefore, by the process of the original nothing's duration approaching the infinite, the original nothing provided opportunity approaching the infinite for the extremely low probability of its change event. Consequently it ultimately occurred, which occurrence was an unavoidable ultimate necessity and which occurrence we call the "Big Bang".

Resolution of the problem of an entire universe appearing and expanding outward through the portal of a dimensionless singularity is fully developed and resolved in *The Origin and Its Meaning* and *How and Why the Universe Began* as is, also, the problem of why the mutually off-setting "halves" did not promptly cancel each other by re-combining.

PART II -- THE EXPANSION OF THE UNIVERSE FROM THE "BIG BANG" OUTWARD

On the basis of the foregoing, the Hubble - Einstein conception of the universe is not valid and the Universal Decay conception of the universe is valid and that validity has already been demonstrated by experiment and observation. The following analysis of the expansion of the universe outward following the first instant of the "Big Bang" is in terms of the Universal Decay conception of the universe as set forth in summary above. The analysis is of the mechanics of the travel of matter outward from its "Big Bang" source [some of it ultimately being we the observers] and of the mechanics of the travel of light from such material sources wherever they are at the time that the light that we later observe is emitted. Results of the analysis are that there is a limit to how far back into the past we are theoretically able to observe [quite regardless of the quality of our observation instrumentation] and that the age of the universe is at least about 30 billion years.

1 - The Travel of Matter and Light

The first step is to develop formulations that describe the travel of the two different traveling entities, light and matter, at various times in the past from at the beginning to the present.

The travel of matter originated at the location of the "Big Bang" singularity and was initially radially outward from that location. While mass cannot travel at light speed the initial speed of the "Big Bang" product particles was sufficiently near the then [initial un-decayed] light speed so as to be taken as such as is developed below. Two effects then proceeded to slow the outward velocities: the decay of the speed of light [the upper limit on particle velocity] and the gravitational slowing [the centrally directed gravitational acceleration, caused by the total mass, decelerating the outward velocities].

The treatment here is of the estimated "average" or "typical" cosmic body [e.g. galaxy], treated as that from its initial form as myriad fundamental particles at the instant of the "Big Bang" -- the particles ultimately destined to form that particular "typical" body, through its form as we know it now. [While not of concern in the present analysis, once the outward travel began the particles experienced local gravitational effects in addition to the overall general slowing -- effects that deflected paths from being purely radially outward and that lead to "clumping" and the formation of structure in the universe.]

The travel of light originated from the above traveling matter, at its various locations and times throughout the universe from the first instant on. It was radially outward from wherever its source was at the time of emission. Its speed was the speed of light at the decayed value for the time after the "Big Bang" that the light was emitted.

a. *The Travel of Light Outward From Astral Sources*

Astral / cosmic source light emitted long ago was emitted at a higher "light speed" than our local contemporary light and continues to travel at that faster speed forever as explained above under the sub-heading of "Universal Decay". On the other hand, the matter originating with the "Big Bang" cannot have traveled at light speed [because its mass would then be infinite] other than nearly so initially before being slowed by gravitation. Therefore, all cosmic source light has been traveling at greater speeds than the cosmic bodies that are home for observers of the light.

Consequently, the most ancient light that it would be theoretically possible for us to observe would be light from a cosmic source that exited the "Big Bang" in the diametrically opposite direction to that of the planetary home [or its components before they became the home planet] of we, the observers. That way, the ancient light has to travel a maximally greater distance from its location where and when emitted to our location where and when we observe it than did our planetary home have to travel from its location when the light was emitted to its location when we observe the light. In other words, ancient light is light that has been traveling a long time and, therefore, has traveled a great distance. The home of we, the observers cannot travel so fast and must, therefore, have a "head start" of distance to be able to arrive at the meeting place of light and observer at the same moment as the faster light. The largest "head start" is the handicapping of the light by placing its source diametrically opposite the location of the observers.

Standard International [SI] units are used; however the great range of magnitudes of the quantities considered calls for their being expressed sometimes in alternative astronomical units: time in *Gyrs = Years*$\cdot 10^9$ rather than *seconds* and distances in *"our" G-Lt-Yrs = 10^9 × [Light Years at our contemporary speed of light]* rather than *meters*. [Note: G-Lt-Yrs is always "our" G-Lt-Yrs.] Those are obtained by the following factors.

(E-5) $\quad k_{time} = 60 \cdot 60 \cdot 24 \cdot 365 \cdot 10^9 \quad$ seconds per giga year $\quad [sec/_{Gyr}]$

$\qquad k_{dist} = k_{time} \cdot [\text{"Our" Light Speed}] \quad$ meters per giga light Year

$\qquad \quad\;\; = k_{time} \cdot [2.997,924,58 \cdot 10^8] \quad [meters/_{G-Lt-Yr}]$

For notes concerning precision see References [4].

For the present the age of the universe is taken to be unknown so that *Age* is a variable. Then based on Equation *E-3*, the original speed of light, *c(0)*, at the instant of the "Big Bang", just before the first moment of the Universal Decay, is obtained as in Equation *E-6*, below.

(E-6) $\quad c(t) = c(0) \cdot \varepsilon^{-t/\tau} \quad meters/_{sec} \qquad$ [light decay per Equation E-3]

$\qquad c(Age) \equiv 2.997,924,58 \cdot 10^8 \quad meters/_{sec} \qquad$ ["our" c, now]

$\qquad c(Age) = c(0) \cdot \varepsilon^{-Age/\tau} \qquad\qquad\qquad$ [set t = Age in Equation E-3]

$\qquad c(0) \;\;\; = c(Age) \cdot \varepsilon^{+Age/\tau} \qquad\qquad\;\;\;$ [solve for c(0)]

$\qquad \qquad\;\;\; = 2.997,924,58 \cdot 10^8 \cdot \varepsilon^{Age/\tau} \quad meters/_{sec}$

Then, the speed of light at any arbitrary time, t, after the "Big Bang" for any arbitrary age of the universe, Age, is as follows.

(E-7) $\quad c(t, Age) = c(0) \cdot \varepsilon^{-t/\tau} = [2{,}997{,}924{,}58 \cdot 10^8 \cdot \varepsilon^{Age/\tau}] \cdot \varepsilon^{-t/\tau} \quad meters/sec$

b. The Travel of Cosmic Bodies Outward From the Origin of the "Big Bang"

To determine the travel of cosmic bodies outward from the "Big Bang" one needs to know the initial velocities and the manner in which they subsequently were reduced by gravitation and other effects. The initial radially outward velocities were so close to the then speed of light as to be that speed for the practical precision here being used. That determination develops as follows.

(1) The Initial Radially Outward Velocities

The universe has existed for billions of years and is still expanding. Therefore, the initial velocity / energy of the "Big Bang" product particles must have been near, if not at or greater than, the escape velocity / energy. The escape velocity / energy for any one particle of the initial "Big Bang" universe is calculated as follows. [The calculation is done non-relativistically here and consequently produces apparent velocities much greater than that of light. They represent velocities nearly at light speed with greatly increased mass.]

Gravitational escape velocity is that velocity the kinetic energy of which just equals in magnitude the potential energy of position in the gravitational field for which the escape velocity is being determined. The non-relativistic escape velocity of a particle develops as follows.

(E-8) $\quad$ **Kinetic Energy** $= \begin{bmatrix} \text{Gravitational} \\ \text{Attraction} \end{bmatrix} \times \begin{bmatrix} \text{Particle Center to Universe} \\ \text{Center Distance} \end{bmatrix}$

Using:
$\quad v_{esc} \equiv$ escape velocity
$\quad m_p \equiv$ mass of the particle
$\quad m_U \equiv$ mass of the Universe [after the initial, essentially instantaneous, mutual annihilations]
$\quad d_0 \equiv$ distance [from the center of mass of the particle to the center of mass of the universe]
$\quad G \equiv$ gravitation constant [un-decayed original value at the time of the "Big Bang"]

Then:

$$\tfrac{1}{2} \cdot m_p \cdot v_{esc}^2 = G \cdot \left[\frac{m_p \cdot m_U}{d_0^2}\right] \times d_0$$

$$v_{esc} = \left[\frac{2G \cdot m_U}{d_0}\right]^{1/2}$$

For that formulation the needed data are: the gravitation constant, G, the mass of the universe, m_U, and the separation distance, d_0. Estimating the Mass of the Universe, m_U, proceeds by estimating the average mass density, ρ, and the volume. The universe mass is then the product of the two. The mass density of the universe, ρ, develops as follows.

Astronomical analyses treat a "critical density" of the universe, ρ_c, which is the particular value of the average density that is on the boundary separating the case of an open (expanding forever) versus closed (eventually gravitationally recontracting) universe. The critical density relates to the escape velocity presented in Equation E-8, above. The development begins with equating kinetic and potential energy in the form of the next to last line of that equation as in Equation $E-9$, below.

$$(E-9) \quad \tfrac{1}{2} \cdot m_p \cdot v^2 = G \cdot \left[\frac{m_p \cdot m_U}{d} \right]$$

The "Hubble Law" states that the velocity of an astral object is proportional to its distance. That law, where H_0 is the "Hubble Constant", is

$$(E-10) \quad v = H_0 \cdot d$$

The total mass inside a sphere of radius d is

$$(E-11) \quad M = [\text{Volume}] \cdot [\text{density}] = [4/3 \cdot \pi \cdot d^3] \cdot [\rho]$$

Substituting in Equation $E-9$ for v with Equation $E-10$ and for m_U with Equation $E-11$ the result is as follows.

$$(E-12) \quad \tfrac{1}{2} \cdot m_p \cdot [H_0 \cdot d]^2 = G \cdot \left[\frac{m_p \cdot [[4/3 \cdot \pi \cdot d^3] \cdot \rho]}{d} \right]$$

$$\rho = \frac{3 \cdot H_0^2}{8 \cdot \pi \cdot G} \qquad \text{[Simplifying and solving for } \rho\text{]}$$

That formulation is intended to give the average density of a portion of the universe of volume $4/3 \cdot \pi \cdot d^3$ such that the mass is on the boundary between escape from that volume and ultimate recapture. It would also, then, be the critical average density, ρ_c, for the overall universe, except for the following problem.

The very concept of the "Hubble Constant" is only valid in terms of the Hubble - Einstein theory that it is space itself that is expanding. It is that which would, if valid, justify the concept of one number, a "universal constant", representing the ratio of distance to velocity. The analogy given for the Hubble - Einstein concept of H_0 is that of the blowing up of a balloon or the rising of a loaf of bread in both of which examples the separation velocity of two locations within is proportional to their separation distance.

However, the "Hubble Constant" and concept are not valid, as already presented. The *form* of Equation $E-12$ is valid and correct, but the constant, H_0, must be replaced with a valid number, the correct ratio of distance to velocity for the object the escape of which is being considered, and that number is not a constant but, rather, depends on the particular circumstances.

Even in Hubble - Einstein terms, the "Hubble Constant", H_0, would better be referred to as the "Hubble Parameter". Not even the first digit of its numerical value is securely determined and its value has been taken to be over a range of from less than $H_0 = 50$ to nearly $H_0 = 100$ for various calculations and estimates by various researchers.

Further, in the "Hubble Law", $v = H_0 \cdot d$, the distance d is the distance of the astral object from the *observer*. The correct distance for the form of Equation $E-12$, that is the distance as in Universal Decay terms not Hubble - Einstein terms, is the distance *outward from the origin* of the "Big Bang". In other words, the "law" is that the object's outward velocity from the origin of the "Big Bang" must be, and must have been, relatively faster if its distance outward, the time-integral of that velocity, is greater, which is obvious. The "Hubble Law" is correct to that extent, but only to that.

Of course the Hubble - Einstein cosmology involves even greater error in attributing redshifts solely to the Doppler Effect of the astral object's velocity rather than the dominant cause, the Universal Decay. That means that determinations of the distance of astral objects by taking their outward velocity from the redshift as a purely Doppler effect, an incorrect velocity, and multiplying it by H_0, an invalid number and concept, can produce only distances in error.

Because for lesser distances from now back into the past [perhaps to 4 or 5 Gyrs ago] Hubble - Einstein redshift calculations of distance deviate relatively less from the correct Universal Decay calculations a look at the results given by Equation E-12 may nevertheless be somewhat helpful in estimating universe average mass density. Depending on the value of H_0 used in Equation E-12, various values for the mass density ρ result, for example:

Value of ρ with the now favored H_0 = 72 km/sec/megaparsec:
$\rho = 9.8 \cdot 10^{-27}$ kg/m3

Value of ρ with the past favored H_0 = 49 km/sec/megaparsec:
$\rho = 4.5 \cdot 10^{-27}$ kg/m3

On the other hand estimates of ρ, rather than theoretical calculations as just above, have been made by estimating the mass of a typical galaxy, that done by estimating the number of stars in a galaxy and multiplying by the estimated average star mass and considering the galaxy's rotational dynamics; then counting the number of galaxies in a volume of space, the process performed for increasingly larger volumes. That procedure has produced a universe mass density estimate of:

Value of ρ from estimating star mass densities:
$\rho \approx 10^{-27}$ kg/m3

Having, then, estimates ranging from about 1 to 10 times 10^{-27}, a reasonable value to use for the mass density of the universe would be the average, about:

(E-13) $\rho_U \approx 5 \cdot 10^{-27}$ kg/m3

Next the volume of the universe is needed so as to obtain the universe's mass as the product of the mass density and the volume. The <u>volume of the universe</u> develops as follows.

The particles of matter of the universe cannot have commenced their travel outward from the origin of the "Big Bang" at one same speed; rather their initial speeds must have been over a range of speeds, which would have produced a wide distribution in space as their travel developed. While the preceding analysis has developed an *average* mass density for the universe, ρ, the actual density must vary substantially even on the scale of large volumes. Therefore, to address the issue of to what volume the average mass density is to be applied requires addressing the issue of the distribution of the initial velocities of the matter emerging from the "Big Bang" because that velocity distribution is the cause of the spatial distribution of astral objects.

The analysis further on below related to Figure E-2c, *First Phase of The Expansion of The Universe -- Velocities for Age = 30 Gyrs Case*, shows that the limits on the range of initial energies of those emerging particles set that range to energies of about 0 to 3,000 × [the escape energy]. Those limits are the obvious lower limit of zero and an upper limit of energy so great that the matter fails to slow to non-relativistic speed ever. However, that is a very large range. Even to only 1,000 is quite large. The range used here for sample cases will be from 1 to 1,000 × [the escape energy]. We cannot know the exact distribution of those energies; but, there are energy distributions of other natural phenomena that can be a guide.

The energy distributions considered are that of Planck Black Body Radiation and of the Maxwell - Boltzman treatment of the kinetic theory of gases. Replacing the case-specific constants [π, h, c, k, 2, and parameter T] with summary case-neutral constants the form of those distributions is as in Equation E-14, and they appear as in Figures E-1a and E-1b, below.

```
(E-14)   Where: F is relative energy [multiple Factor of escape energy].
                n(F) is the number of particles of energy multiple F.
                p(F) is the probability of interval [F+ΔF], ΔF→0.
```

Planck:
$$n(F) = \frac{K1 \cdot F^5}{\varepsilon^{K2 \cdot F} - 1}$$

Maxwell-Boltzman:
$$p(F) = \frac{K3 \cdot F^{1/2}}{\varepsilon^{K4 \cdot F}}$$

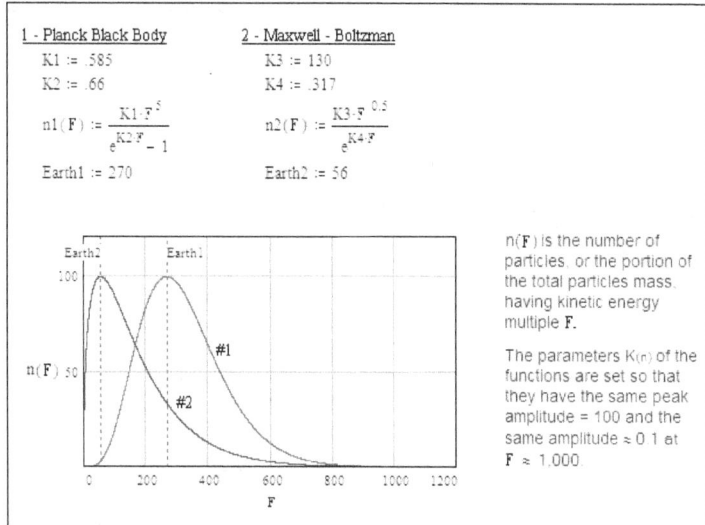

Figure E-1a -- "Big Bang": Some Theoretical Rate Distributions of Initial Particle Energies

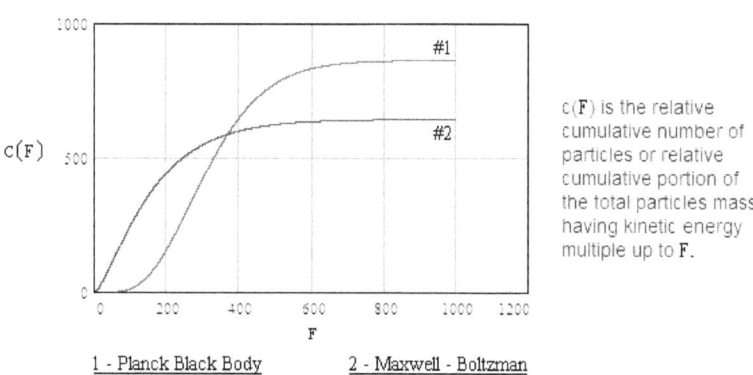

Figure E-1b -- Cumulative Distributions for Figure E-1a

Where we observers on planet Earth fall in the distributions in the above Figure E-1a should be considered. It can only be presumed that Earth is not unusual with regard to its component particles' initial velocities, which would call for placing it at a distribution peak. But, there are two choices shown, likely neither is exact, and Earth is not necessarily so usual as to fall exactly at the peak of any distribution. Form #2 requires less total energy and also is chosen because otherwise the resulting velocities at `Age = 30 Gyrs` would appear to be in error relative to the known velocity of Earth (see before Equation `E-27`, further below. Calculating for the two cases in Figure E-1a shows that the variation in choices produces little variation in the overall results and in the age, `30 Gyrs`, of the universe and the theoretical limit, about `27 Gyrs`, on how far back in the past can be observed. The Earth case parameter, `F = 55`, [see Tables 2a and 2b, further below] is chosen to set it at the distribution Form #2 peak.

From the above data the summary conclusions in Table *E*-1c, below, can be drawn.

Form #1 -- Planck Black Body Radiation			
Percent of Maximum Amplitude, `n(F)`:	`100 %`	`95 %`	`90 %`
Range in Distribution is `F = 0 to`:	`1,000`	`565`	`500`
That Range as Percent of the Maximum	`100 %`	`57 %`	`50 %`
Resulting Indicated Universe Radius	`~14 G-Lt-Yrs`	`~14 G-Lt-Yrs`	`~14 G-Lt-Yrs`
Form #2 -- Maxwell - Boltzman Gas Kinetic Theory			
Percent of Maximum Amplitude, `n(F)`:	`100 %`	`95 %`	`90 %`
Range in Distribution is `F = 0 to`:	`1,000`	`455`	`360`
That Range as Percent of the Maximum	`100 %`	`46 %`	`36 %`
Resulting Indicated Universe Radius	`~14 G-Lt-Yrs`	`~14 G-Lt-Yrs`	`~14 G-Lt-Yrs`

Table E-1c -- "Big Bang" Initial Energy Distributions Summary Data Conclusions

In the above table, the "*Resulting Indicated Universe Radius*" in `G-Lt-Yrs` is obtained as follows. Figure E-4d, further on below, *Second Phase of The Expansion of The Universe -- Distances for Age = 30 Gyrs Case* indicates a present [at `Age of the universe ≈ 30 Gyrs`] radius of the matter-containing volume of the universe as about `8 G-Lt-Yrs`. However, the radius applicable to the above obtained universe mass density should be based on an earlier time because the investigations into estimating that density had to treat astral objects which we observe as they were some time in the past: their distance from us divided by the speed of their light. Taking those earlier times as having been in the range of `0 to 7-8 Gyrs` into the past, which corresponds to volumes in the ratio to each other of the cube of those distances as `[0, 1, 8, 27, 64, 125, 216, 343, 512]`, and cumulatively in ratio as `[0, 1, 9, 36, 100, 225, 441, 784, 1296]` then it is reasonable to take the applicable universe radius as that which existed at the time into the past corresponding to about half the maximum cumulative volume, `t ≈ 6.5 Gyrs` ago. Figure E-4d indicates the radii given in the above Table *E*-1c for the related table columns at that time ago, `≈ 14 G-Lt-Yrs`.

Then, the estimated radius of the universe for the present calculation is:

(E-15) $R_U = 14$ `G-Lt-Yrs` $= 11 \cdot 10^{24}$ meters.

Therefore the mass of the universe, as the product of its volume based on that radius and its Equation `E-13` density, is:

(E-16) $m_U = 3 \cdot 10^{49}$ kg.

[Calculating with alternative values for the mass of the universe ranging from 10^{46} to 10^{53} kg produces no significant change in the general results developed below as can be verified using the forms of the calculations presented further on below. That is, while velocities and distances vary somewhat, the necessary age of the universe remains at about the *30 Gyrs* and the maximum distance back into the past that it is theoretically possible to observe remains at about the *27 Gyrs* developed further on below.]

[A possible concern over circular cause and effect reasoning here is not valid. The results presented are based on numerous iterations of calculations over a range of complexly interacting variables.]

With the mass of the universe now resolved the other quantities needed to calculate the escape velocity of the universe can be addressed. The <u>Separation Distance</u>, d_0, is the radius of the universe at the moment that expansion began being at a rate consistent with the long term development of the universe as compared to its initial more rapid [essentially instantaneous] development commonly referred to as "inflation". That value is $d_0 = 4.0 \cdot 10^7$ *meters*.

$G(0)$, the <u>Gravitation Constant</u> at its original un-decayed value at the time of the "Big Bang" is as follows.

(E-17) $G(t) = G(0) \cdot \varepsilon^{-3 \cdot t/\tau}$ $m^3/kg\text{-}s^2$ [Form per the Universal Decay and the $[m^3]$ requires $\tau \to \tau/3$]

$G(Age) = 6.672,59 \cdot 10^{-11}$ $m^3/kg \cdot s^2$ ["Our" G, now]

$G(Age) = G(0) \cdot \varepsilon^{-3 \cdot Age/\tau}$ [Set t = Age in G(t)]

$G(0) = G(Age) \cdot \varepsilon^{+3 \cdot Age/\tau}$ [Solve for G(0)]

$\qquad = 6.672,59 \cdot 10^{-11} \cdot \varepsilon^{3 \cdot Age/\tau}$ $m^3/kg \cdot s^2$

Then, the gravitation constant at any arbitrary time, t, after the "Big Bang" for any arbitrary age of the universe, *Age*, is as follows.

(E-18) $G(t, Age) = G(0) \cdot \varepsilon^{-3 \cdot t/\tau}$

$\qquad = [6.672,59 \cdot 10^{-11} \cdot \varepsilon^{3 \cdot Age/\tau}] \cdot \varepsilon^{-3 \cdot t/\tau}$ $meters/sec$

Two values for the *Age* of the universe are addressed in this analysis to present the thesis and its validation. The currently accepted values in Hubble - Einstein cosmology range *Age = 13.5 to 14.7 Gyrs*. Representing those *14.0 Gyrs* will be used. As developed below, the present analysis indicates that *Age = 30.0 Gyrs*. Then, using $\tau = 11.3373$ *Gyrs* from Equation *E-2* the following values for $G(0)$ result.

(E-19) <u>For Age = 14 Gyrs</u>   <u>For Age = 30 Gyrs</u>

$G(0) = 2.711 \cdot 10^{-9}$ $G(0) = 1.870 \cdot 10^{-7}$

The escape velocity per Equation *E-8* for those cases of age of the universe are:

<u>For Age 14 Gyrs</u>   <u>For Age 30 Gyrs</u>

$v_{esc} = 6.4 \cdot 10^{16}$ m/s $v_{esc} = 5.3 \cdot 10^{17}$ m/s

Those values are so large relative to the speed of light at the time of the "Big Bang",

(E-20)	For Age 14 Gyrs	For Age 30 Gyrs
	$c(0) = 1.031 \cdot 10^9 \text{ m}/\text{s}$	$c(0) = 4.227 \cdot 10^9 \text{ m}/\text{s}$

that it is certain that the initial particle velocities, at the time of the "Big Bang", were very nearly the then speed of light. That is, the initial particle velocities could not be, nor exceed, light speed as the non-relativistically calculated escape velocities of Equation E-20 call for. The accommodation to relativity means that the actual speeds were very near light speed and the masses were significantly relativistically increased.

As noted earlier, to determine the travel of cosmic bodies outward from the "Big Bang" one needs to know, first, the initial velocities and then the subsequent manner of the reduction in the cosmic bodies' velocities by gravitation and other effects. The initial velocities have been found to be essentially the value of the speed of light at the time of the "Big Bang". At that point two effects proceeded to slow the outward velocities: the decay of the speed of light [the upper limit on particle velocity] and the gravitational slowing [the centrally directed gravitational acceleration caused by the total mass].

(2) The Progressive Reduction in the Cosmic Bodies' Initial Velocities

The overall process must be divided into two phases:

- First, the relativistic phase during which the speed is continuously almost that of light and the effect of gravitation is dominantly not a reducing of the speed but a reducing of the amount that the mass has been relativistically increased, and

- Second, the non-relativistic phase during which the mass, now reduced to essentially rest mass, remains essentially the same and the dominant effect of gravitation is to reduce the speed.

Of course the change from the first to the second phase is not sharp, but rather a gradual smooth transition. For the purposes of these calculations, however, the choosing of a specific transition point [hereafter termed the `ChangePoint`] is needed. That point is determined as follows.

(2a) The First Calculation Phase -- Speed ≈ Light Speed

The first phase calculation is in terms of energy, the gradual transfer of initial kinetic energy into gravitational potential energy. Energy calculations in themselves are not relativistic. The kinetic energy speed will be treated non-relativistically, that is, the mass is taken as at its rest value and the kinetic energy then is taken as all residing in the [theoretical] velocity squared, that theoretical velocity not constrained by a speed of light limitation. Then, the calculated effect of gravitation, of the transfer of kinetic energy into gravitational potential energy, appears as a gradual reduction of that theoretical velocity. When that theoretical velocity has been reduced by gravitation down to the actual [at that time as decayed] light speed then the `ChangePoint` from the relativistic to the non-relativistic treatment has been reached.

During that first phase the distance component of the gravitational potential energy calculation is readily available as the time integral of the known speed, the speed of light. The velocity as a function of time then develops as follows.

E – THE 'BIG BANG' OUTWARD COSMIC EXPANSION

The first phase distance, $d(t, Age)$, traveled outward from the "Big Bang" source location, as a function of time is the time integral of the velocity as Equation $E-21$, below, which is based on Equation $E-7$, above [and includes the initial separation distance, $d_0 = 4.0 \cdot 10^7$ meters, of the earlier above calculation of the mass of the universe, which distance is negligible, however].

(E-21)
$$d(t, Age) = d_0 + \int_0^t c(t) \cdot dt$$

$$= 4.0 \cdot 10^7 + \int_0^t \left[[2.997{,}924{,}58 \cdot 10^8 \cdot \varepsilon^{Age/\tau}] \cdot \varepsilon^{-t/\tau} \right] \cdot dt$$

The velocity as a function of time, $v(t, Age)$, is, starting from Equation $E-8$:

(E-22)
$$v_{esc} = \left[\frac{2G \cdot m_U}{d_0} \right]^{\frac{1}{2}}$$

so that:

$$v(t, Age) = \left[\frac{2G \cdot m_U}{d(t, Age)} \right]^{\frac{1}{2}}$$

As pointed out during the evaluation of the mass of the universe earlier above, the matter of the universe moved outward from the "Big Bang" at a wide range of speeds. Those various speeds resulted, of course, from the particles of matter having various initial velocities / energies which, as presented just before and in conjunction with Figures E-1a and E-1b, are to be sampled over the range $F = 1$ to $1{,}000$ X [the escape energy]. That range is incorporated into the formulation by the multiple factor, F, included in the final expression for the first phase $v1(t, Age)$ per below.

(E-23)
$$v1(t, Age) = \left[\frac{F \cdot 2G \cdot m_U}{d(t, Age)} \right]^{\frac{1}{2}} \quad \text{[1st Phase Matter Velocities]}$$

The decaying speed of light is per Equation $E-7$, repeated below,

(E-24) $\quad c(t, Age) = [2.997{,}924{,}58 \cdot 10^8 \cdot \varepsilon^{Age/\tau}] \cdot \varepsilon^{-t/\tau}$

and the value of time, t, producing $\boxed{v1(t, Age) \equiv c(t, Age)}$ is the sought *ChangePoint* for the particular initial velocity / energy multiple factor, F, and Age, the end of calculation for the first, the relativistic, phase.

Tables *E-2a* and *E-2b*, below summarize the results for the first phase for both $Age = 30$ and $Age = 14$ *Gyrs*, and the results are also presented graphically for $Age = 30$ *Gyrs* in Figure E-2c, following the tables.

For: Universe Age = 30 Gyrs, which means that:
Initial Light Speed = $4.226,895,62 \cdot 10^9$ m/s
Initial Gravitation Constant, G = $1.870,24 \cdot 10^{-7}$ m^3/kg-s^2

F-Factor	"ChangePoint" At Time[Gyrs]	At Velocity[m/s]	Distance From Origin* ChangePoint	Now, Age	Relative! Abundance
1	0.004713	$4.225 \cdot 10^9$	0.066	0.005	22.
3	0.01414	$4.222 \cdot 10^9$	0.199	0.014	37.
10	0.04721	$4.209 \cdot 10^9$	0.661	0.047	63.
32	0.1518	$4.171 \cdot 10^9$	2.097	0.152	93.
55 Earth	0.2621	$4.131 \cdot 10^9$	3.568	0.262	100.
100	0.4812	$4.052 \cdot 10^9$	6.364	0.481	90.
316	1.5960	$3.672 \cdot 10^9$	18.222	1.596	23.
1000	6.0840	$2.472 \cdot 10^9$	38.788	6.084	0.1
≈3000	→ ∞	c(t)	→ ∞	→ ∞	

* = Decayed to Change Point, Age; G-Lt-Yrs.   ! = Estimate per Figure E-1a

Table E-2a - The Universe's First Phase of Expansion, Age=30 Gyrs Case -- The Relativistic Phase at (Essentially) Light Speed, From t = 0 to t = ChangePoint

For: Universe Age = 14 Gyrs, which means that:
Initial Light Speed = $1.030,357,62 \cdot 10^9$ m/s
Initial Gravitation Constant, G = $2.711,29 \cdot 10^{-9}$ m^3/kg-s^2

F-Factor	"ChangePoint" At Time[Gyrs]	At Velocity[m/s]	Distance From Origin* ChangePoint	Now, Age	Relative! Abundance
1	0.004715	$1.030 \cdot 10^9$	0.016	0.005	22.
3	0.01416	$1.029 \cdot 10^9$	0.049	0.014	37.
10	0.04721	$1.026 \cdot 10^9$	0.161	0.047	63.
32	0.1518	$1.017 \cdot 10^9$	0.511	0.151	93.
55 Earth	0.2621	$1.007 \cdot 10^9$	0.870	0.259	100.
100	0.48125	$9.878 \cdot 10^8$	1.551	0.471	90.
316	1.5961	$8.953 \cdot 10^8$	4.443	1.488	23.
1000	6.089	$6.024 \cdot 10^8$	9.460	4.708	0.1
≈3000	→ ∞	c(t)	→ ∞	→ ∞	

* = Decayed to Change Point, Age; G-Lt-Yrs.   ! = Estimate per Figure E-1a

Table E-2b - The Universe's First Phase of Expansion, Age=14 Gyrs Case -- The Relativistic Phase at (Essentially) Light Speed, From t = 0 to t = ChangePoint
Figure E-2c

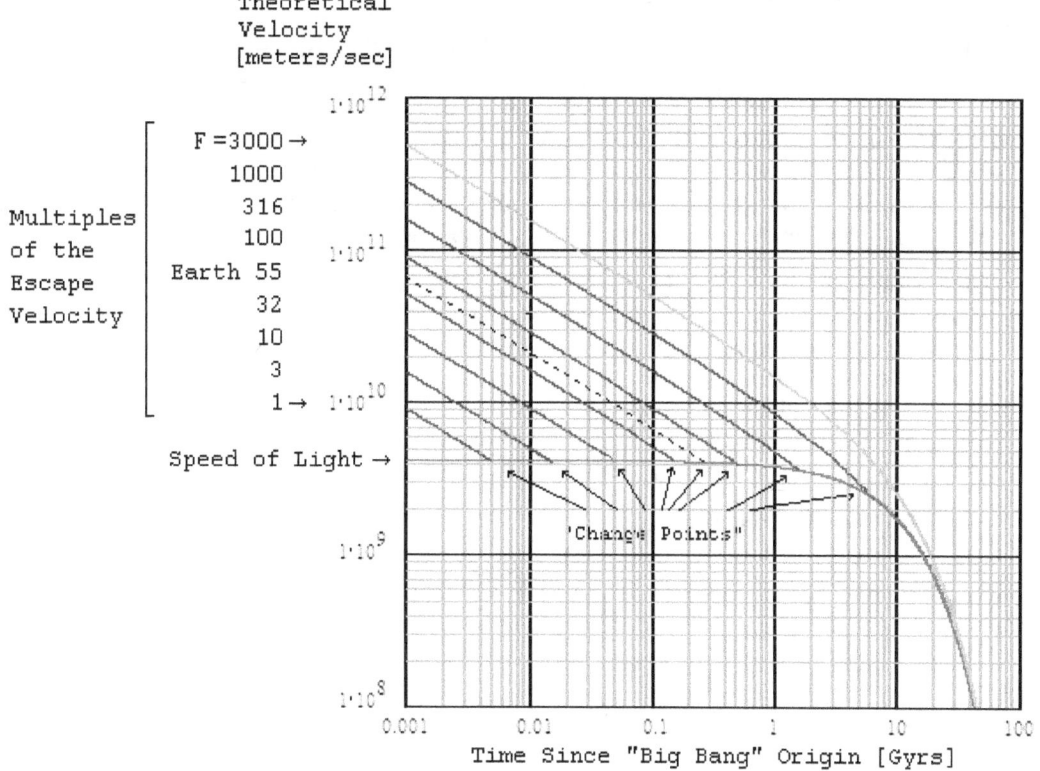

First Phase of The Expansion of The Universe -- Velocities for Age = 30 Gyrs Case

Note that for values of the *F Factor* at about $F = 3,000$ and above the "*ChangePoint*" is never reached because of the decay in the speed of light. For those values the outward moving matter never slows below, essentially, the then on-going decaying light speed.

Note, also, that the large Doppler red shift resulting from v nearly equaling c combined with large decay redshift due to lack of much decay because the time lies only shortly after $t = 0$ results in a redshift relative to our "normal" local wavelengths by a factor of $24 - 28$. The least wavelength of visible light is about 0.38 *microns*. That shifted to $0.38 \times [24 - 28] = [9 - 11]$ *microns*, lies well into the infra-red portion of the spectrum.

Consequently, the light emitted from sources before they reach their "ChangePoints", which light would otherwise lie in the "visible light" portion of the spectrum, lies shifted sufficiently into the infra-red that its detection is relatively unlikely, especially since sources with relatively later "ChangePoints" [more recent, therefore more susceptible to observation] are of relatively small relative abundance. In other words, light emitted from astral sources before they reached their *ChangePoint* is much less likely to be observed.

(2b) The Second Calculation Phase -- Speed < Light Speed

Gravitational slowing is an awkward problem. The amount of gravitational slowing depends on the distance outward from the origin of the "Big Bang"; those distances depend on the velocity function during the travel from the origin outward; and that velocity function depends on the gravitational slowing -- a problem of circular cause and effect. The calculation breaks down into two different modes of behavior because of relativistic effects.

The first phase of the outward expansion, already analyzed above, takes place at essentially the actual speed of light regardless of gravitation. That is because the [theoretical non-relativistic] escape velocities are so large. The kinetic energy essentially resides in the relativistically increased mass until gravitation has reduced the [theoretical non-relativistic] greater than light speed down to passing through and below the actual light speed. During that first phase distances are known because the velocities are known independently of the distance; they are essentially the then current, as decayed, light speeds.

The second phase begins at the end of that first phase's known outward travel to its "*ChangePoint*". At that point the circular cause and effect awkwardness of the problem returns. The inverse square gravitational behavior calls for the current total outward distance squared in its denominator and that depends on the velocity history which depends on the distance history which depends on the velocity history. The solution is to use an approximating function of similar form but not involving velocity. That function can then be adjusted by calibrating it to the known velocity of the Earth now, as developed further below.

The <u>approximating function</u> develops as follows. The precise behavior of the universe's matter expanding outward from the "Big Bang" is as Equation *E-25*, the distance represented by the variable s to avoid confusion with the symbols for differentiation.

(*E-25*)
$$\text{Gravitational Deceleration} = \frac{d^2 s}{dt^2} = - \frac{G \cdot \text{UniverseMass}}{s^2}$$

The general form of the solution to that equation is as Equation *E-26*

(*E-26*)
$$\text{Velocity} = \frac{ds}{dt} = \frac{1}{A \cdot \varepsilon^{B \cdot s} + C}$$

which states that the velocity is inversely proportional to the exponential of the distance, s, as $1/\varepsilon^s$. The procedure will be to use as the approximation to the actual exact velocity function the function: the speed of light, c(t), per Equation *E-24*, multiplied by a factor based on $1/\varepsilon^s$.

However, the velocity function must be in terms of time, t, as the independent variable, not distance, s, else the problem of circular cause and effect remains. Using t instead of s, that is representing the actual exact velocity function with a function multiplying the speed of light, c(t), by a factor based on $1/\varepsilon^t$ resolves that problem but is less accurate an approximation. The problem of accuracy is addressed by calibrating to the known velocity of our planet Earth at time the *Age* of the universe.

Doppler analysis of the cosmic microwave background radiation shows that the absolute velocity of the Earth [absolute relative to the location of the "Big Bang" origin] is now about $3.7 \cdot 10^5$ *meters*/*sec*. The calibration of the velocity approximating function must be such as to produce approximately that velocity now, at time t = *Age* after the "Big Bang"; but, for what case, what value of the *F-Factor*? That issue has been already addressed above and the result is that, for *Earth F* = 55 will be used.

So that the beginning of the second phase will match smoothly with the ending of the first phase the adjustment, $1/\varepsilon^t$, which increases in its effect as t increases, must be formulated to produce zero change when the time is t = ChangePoint. The resulting expression for the second phase velocity $v2$ is Equation E-27, below.

(E-27)
$$v2(t, \text{Age}) = c(t, \text{Age}) \cdot \frac{1}{\varepsilon^{A \cdot (t - \text{ChangePoint})}} \text{ meters/sec}$$

[where A is a constant of value yet to be determined, chosen to calibrate the function, and t must always be t ≥ ChangePoint].

The calibrating constant, A, for the case of *Earth*, F-Factor = 55 is set to the value that produces a velocity at Age of $3.7 \cdot 10^5$ meters/sec, the known value. The A for the highest F-Factor case is set to produce a velocity of the current speed of light, $3 \cdot 10^8$ meters/sec, at Age. The A for each of the other cases is interpolated using a decaying exponential form to provide a general representative set of samples. That form is per Equation E-28 and as depicted in Figure E-3 below it.

(E-28) $\quad \text{Velocity}(i) = 5.452 \cdot 10^9 \cdot [\varepsilon^{-2.2127(9-i)}]$

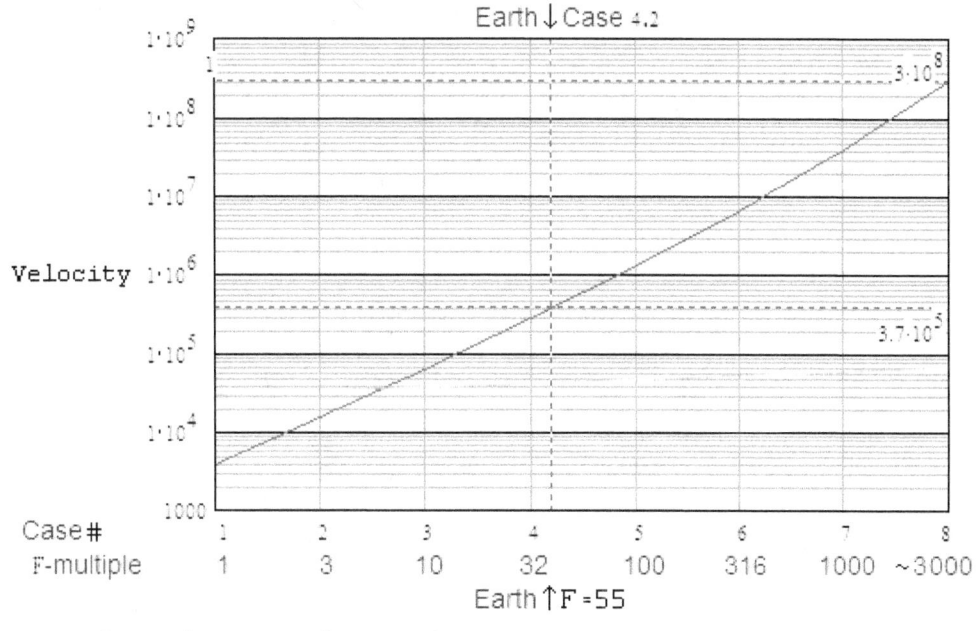

where i = case # = 1, 2, ... 8

Figure E-3
Calibrated Velocities at "Age" for Sample Expansion Cases #1 - #7, and #Earth

The development of the formulations for the distances is presented further below at Equations E-31A through E-33B. Tables 4a and 4b below summarize the results for the second phase for Age = 30 Gyrs and for Age = 14 Gyrs, and the results are also presented graphically for Age = 30 Gyrs in Figure E-4c, *Second Phase of The Expansion of The Universe -- Velocities for Age = 30 Gyrs Case*, and Figure E-4d, *Second Phase of The Expansion of The Universe -- Distances for Age = 30 Gyrs Case*, on the page following the tables.

For: **Universe Age = 30 Gyrs**, which means that:
Initial Light Speed = $4.226,895,62 \cdot 10^9$ m/s
Initial Gravitation Constant, $G = 1.870,24 \cdot 10^{-7}$ m³/kg-s²

F-Factor	ChangePoint Time [Gyrs]	2nd Phase Constant-A	At Age = 30 Gyrs, [Data*] Velocity [m/s]	Distance [G-Lt-Yrs]
1	0.004713	16.92385	$0.00003814 \cdot 10^8$	2.157
3	0.01414	16.98010	$0.0001505 \cdot 10^8$	2.400
10	0.04721	17.04691	$0.0006230 \cdot 10^8$	2.727
32	0.1518	17.12856	$0.002731 \cdot 10^8$	3.205
55	0.2621	17.14622	$0.003700 \cdot 10^8$	3.374
100	0.4812	17.23245	$0.01287 \cdot 10^8$	3.980
316	1.5960	17.37183	$0.06673 \cdot 10^8$	5.388
1000	6.0840	17.57200	$0.3974 \cdot 10^8$	8.034
≈3000	→ ∞	n/a	$2.99792458 \cdot 10^8$	10.526

* = Decayed to Age

Table E-4a - For Age = 30 Gyrs
Summary Data Results for the Universe's Second Phase of Expansion --
The Non-Relativistic Phase From t = "ChangePoint" Onward

For: **Universe Age = 14 Gyrs**, which means that:
Initial Light Speed = $1.030,357,62 \cdot 10^9$ m/s
Initial Gravitation Constant, $G = 2.711,29 \cdot 10^{-9}$ m³/kg-s²

F-Factor	ChangePoint Time [Gyrs]	2nd Phase Constant-A	At Age = 14 Gyrs, [Data*] Velocity [m/s]	Distance [G-Lt-Yrs]
1	0.004715	16.59278	$0.00003814 \cdot 10^8$	1.122
3	0.01416	16.64887	$0.0001505 \cdot 10^8$	1.268
10	0.04721	16.71513	$0.000623 \cdot 10^8$	1.477
32	0.1518	16.79505	$0.002731 \cdot 10^8$	1.811
55	0.2621	16.81082	$0.003700 \cdot 10^8$	1.954
100	0.4812	16.89330	$0.01287 \cdot 10^8$	2.417
316	1.5961	17.01200	$0.06673 \cdot 10^8$	3.669
1000	6.0890	17.09155	$0.3974 \cdot 10^8$	6.295
≈3000	→ ∞	n/a	$2.99792458 \cdot 10^8$	8.034

* = Decayed to Age

Table E-4b - For Age = 14 Gyrs
Summary Data Results for the Universe's Second Phase of Expansion --

E – THE 'BIG BANG' OUTWARD COSMIC EXPANSION

The Non-Relativistic Phase From t = "ChangePoint" Onward

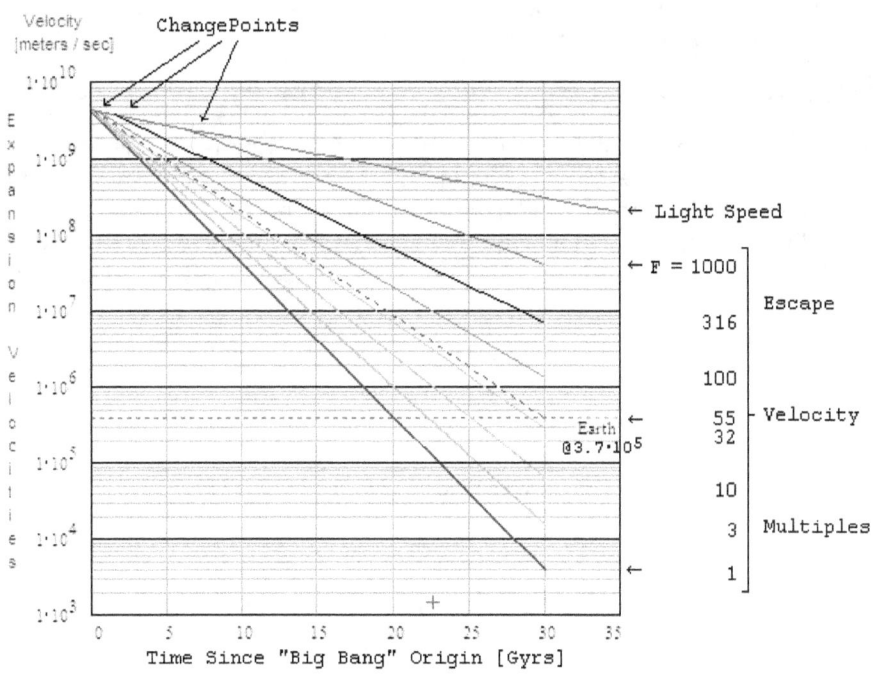

Figure E-4c
Second Phase of The Expansion of The Universe -- Velocities for Age = 30 Gyrs Case

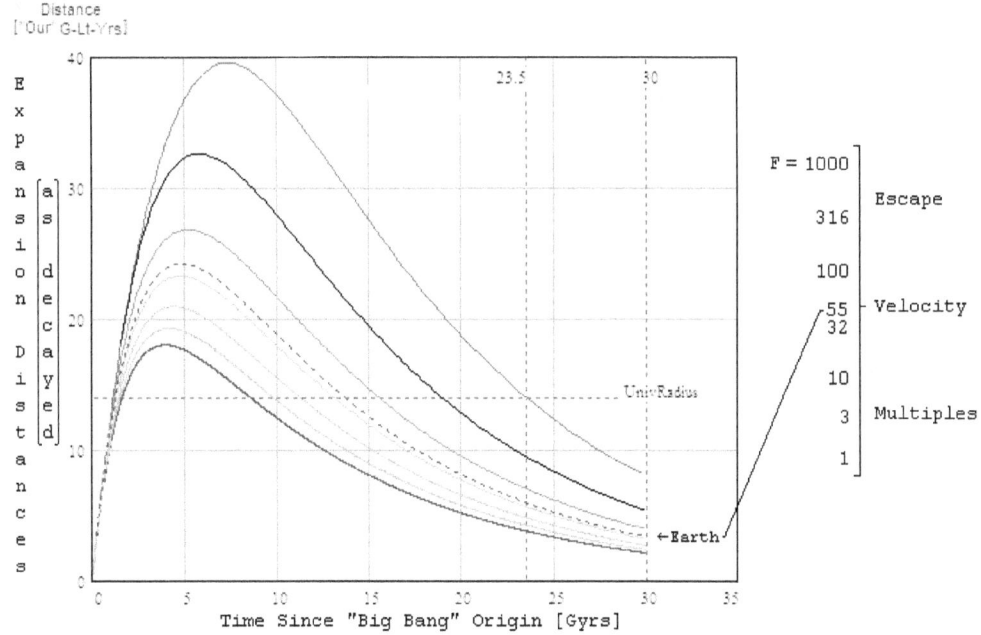

Figure E-4d
Second Phase of The Expansion of The Universe -- Distances for Age = 30 Gyrs Case

[The notation "UnivRadius" in the above Figure E-4d refers to the discussion following Table *E-1c* concerning developing the appropriate volume of the universe to use in conjunction with the universe's density to obtain the universe's mass.]

2 - *Observing, from Our Planet Earth, Light Emitted by Astral Sources*

Now that formulations have been developed that describe the travel of the two different traveling entities, light and matter, at various times in the past from at the beginning to the present the problem of observing light from astral sources can be addressed. The problem is to determine under what conditions light emitted long ago will have traveled to the exact present location of an observer that has also been performing its own travel while the light to be observed has been traveling.

For convenience the following quantities are defined.

```
Age  ≡ Age now of the universe = time "now".
Back ≡ how long ago it is theoretically possible to observe.
Then ≡ the Age of the universe then [Back ago].
     = Age - Back.
```

The distance that the light travels [continuously at whatever its speed was when it was first emitted from its astral source] from when first emitted at time t until now, at time *Age*, is:

(E-29) LightTravels(t,Age) = [Speed(t)]·[Travel Time]

= [c(t,Age) [m/s]]·[[Age - t] [s]]

= c(t,Age)·[Age - t]   meters

Consistent with the mnemonic terminology above, *LightTravels(t,Age)*, the motion of the cosmic bodies involved will be termed as in Equation *E-30*.

(E-30) WhereUs(t,Age,F)   ≡ location of we observers

WhereSource(t,Age,F) ≡ location of source of the observed light

≈ -WhereUs(t,Age,F)

[to account for the formulations being of similar form except that that their travels are diametrically opposite for the purposes of the present calculations and their "F" values can be different].

Taking the location of the "Big Bang" singularity as at distance zero and noting that the initial distance, $d_0 = 4.0 \cdot 10^7$ *meters*, is negligible [less than one "our light years" on a scale of "giga our light years"], then the formulation for *WhereUs(t,Age)* develops as follows.

(E-31)   Travel from the "Big Bang" outward

(E-31A) Travel during time t=0 to time t=ChangePoint:

WhereUsA(t,Age,F) = [Travel of Matter in 1st Phase]

$$= \int_0^t [\text{Speed of Matter in 1st Phase}(t,Age,F)] \cdot dt$$

$$= \int_0^t [\text{Decaying Light Speed}(t,Age)] \cdot dt$$

E – THE 'BIG BANG' OUTWARD COSMIC EXPANSION

$$= \int_0^t c(t, Age) \cdot dt \quad \text{meters}$$

(E-31B) Travel from time t=ChangePoint onward:

   WhereUsB(t, Age, F) =

   = [1st Phase Travel to ChangePoint] +
       + [2nd Phase Travel(t, Age, F)]

   = [WhereUsA(ChangePoint, Age)] +
       + [2nd Phase Travel(t, Age, A)]   [F → A]

$$= [\text{1st Term}] + \int_{\text{ChangePoint}}^t [\text{DecayingLightSpeed}] \cdot [\text{GravitySlowing}] \cdot dt$$

$$= [\text{1st Term}] + \int_{\text{ChangePoint}}^t [c(t)] \cdot \frac{1}{\varepsilon^{A \cdot [t - \text{ChangePoint}]}} \cdot dt \quad \text{meters}$$

and where "A" is the "2nd Phase Constant-A" of Tables 4a and 4b.

However, there is one more factor in the development of *WhereUs(t,Age)*, the general effect of the Universal Decay. The Universal Decay produces an acceleration on all bodies which acceleration is: centrally directed, independent of separation distances, and the same and constant everywhere except that the amount of that acceleration also exponentially decays. Its value now is *(8.74 ± 1.33) × 10⁻⁸ cm/s2*. That acceleration produces a gradual contraction of the overall universe; that is, the exponential decay of the length *[L]* aspect of all quantities is also a decay of the distance spacings in the universe.

The acceleration is evidenced by galactic rotation curves and by the travel of the Pioneer 10 and 11 space craft. That the Universal Decay is not directly observable because our measuring equipment [our "ruler"] is also decaying, as noted earlier above, prevents our direct observation of the contraction in the case of our solar system and that of galactic rotation curves. However, in the case of the Pioneer 10 and 11 satellites and the case of galactic rotation curves, the decay has been detected because it forces orbital / path behavior that would not be present if there were no decay, and that behavior can be and has been observed -- e.g. the Pioneer space craft are not as far outward from the Sun as they should be were there no Universal Decay contraction.

Consequently, as the various bodies in the universe travel outward from the location of the "Big Bang", the distances that they have already traveled continuously decay. And, consequently, the distances traveled by light emitted from the various sources in the universe continuously decay after having been first traveled "undecayed". Therefore, the final form for *LightTravels* is then Equation *E-32*, below [continued from Equation *E-29*]. And, the final form for *WhereUs(t,Age)* is Equations *E-33A* and *E-33B*, below [continued from Equations *E-31A* and *E-31B*, above].

(E-32)   Distance traveled outward from its source until now, time "Age", by light emitted at time "t":

   LightTravels(t, Age) = [Speed] · [Travel Time] · *[As Decayed]*

$$= c(t, Age) \cdot [Age - t] \cdot \varepsilon^{-[(Age - t)/\tau]} \quad \text{meters}$$

PRIME OBJECTIVE TIME, PRIME OBJECTIVE SPACE

(E-33) Distance traveled outward from the "Big Bang" until time "t" by matter originating at the "Big Bang" [t=0, distance=0]:

WhereUs(t,Age,F):

(E-33A): Travel during time t=0 to time t=ChangePoint:

WhereUsA(t,Age,F) = [Travel of Matter in 1st Phase] · *[As Decayed]*

$$= \left[\int_0^t c(t,Age) \cdot dt \right] \cdot \varepsilon^{-[t/\tau]} \quad \text{meters}$$

(E-33B): Travel from time t=ChangePoint onward:

WhereUsB(t,Age,F) =

= [[1st Phase Travel to ChangePoint]·[not decayed] +
+ [2nd Phase Travel(t,Age,A)]] · *[All as Decayed]*

$$= \left[[\text{1st Term}] + \int_{\text{ChangePoint}}^t [c(t)] \cdot \frac{1}{\varepsilon^{A \cdot [t - \text{ChangePoint}]}} \cdot dt \right] \cdot \varepsilon^{-[t/\tau]} \quad \text{meters}$$

a. Calculation of the Maximum Distance into the Past That is Observable

The extreme case of observing ancient light is the observing of light that originated diametrically opposite from us, the observers, relative to the origin of the "Big Bang". That light <u>must</u> travel the distance from its source back to the location of the origin of the "Big Bang" and then further outward to the location of us, the observers. The light originates from its source at time $t = Then$ and is observed by us at time $t = Age$. That distance, for any age of the universe, *Age*, is as follows.

(E-34) LightMustTravel(t,Age,F) =
= -WhereSource(Then,Age,F) + WhereUs(Age,Age,F)

As compared to the above requirement, the actual distance that that light <u>does</u> travel is given by Equation *E-32* with $t = Then$, as follows.

(E-35) $\text{LightDoesTravel}(t, Age) = c(Then, Age) \cdot [Age - Then] \cdot \varepsilon^{-[(Age-Then)/\tau]}$

$= c(Then, Age) \cdot [Back] \cdot \varepsilon^{-[(Back)/\tau]} \quad \text{meters}$

For the light to be theoretically observable by us the above two must be the same.

(E-36) LightMustTravel(t,Age,F) = LightDoesTravel(t,Age)

The only variables in Equation *E-36* [for a particular *Age* and energy multiple, *F*,] are *Back* and *Then*, either of which determines the other per *Then = Age - Back*. The solution to Equation *E-36* is obtained using a computer assisted design program ["Mathcad" in this case]. The applicable form of *WhereUs(t,Age,F)* must be used, *WhereUsA(t,Age,F)* or *WhereUsB(t,Age,F)* depending on the value of *Then* relative to the *ChangePoint*.

The results are presented in Tables *E-5a* and *E-5b*, below.

```
For:  Universe Age = 30 Gyrs, which means that:
      Initial Light Speed = 4.226,895,62·10⁹ m/s
      Initial Gravitation Constant, G = 1.870,24·10⁻⁷ m³/kg-s²
```

F-Factor Abundance	ChangePoint Time [Gyrs]	2nd Phase Constant-A	Observable Past Distance Back [Gyrs]	Then [Gyrs]	Relative!
1	0.004713	16.92385	7.5860	22.4140	22.
3	0.01414	16.98010	8.4081	21.5919	37.
10	0.04721	17.04691	9.8790	20.1210	63.
32	0.1518	17.12856	26.0791	3.9209	93.
100	0.4812	17.23245	27.01869	2.98131	90.
316	1.5960	17.37183	27.56523	2.43477	23.
1000	6.0840	17.57200	27.61821	2.38179	0.1

! = Estimate per Figure E-1a, where Earth, F = 55, is of Abundance 100

Table E-5a
Distance into the Past That is Observable, Age = 30 Gyrs

```
For:  Universe Age = 14 Gyrs, which means that:
      Initial Light Speed = 1.030,357,62·10⁹ m/s
      Initial Gravitation Constant, G = 2.711,29·10⁻⁹ m³/kg-s²
```

F-Factor Abundance	ChangePoint Time [Gyrs]	2nd Phase Constant-A	Observable Past Distance Back [Gyrs]	Then [Gyrs]	Relative!
1	0.004715	16.59278	3.4828	10.5172	22.
3	0.01416	16.64877	3.7136	10.2864	37.
10	0.04721	16.71513	4.0702	9.9298	63.
32	0.1518	16.79505	4.6849	9.3151	93.
100	0.48125	16.89330	5.9772	8.0228	90.
316	1.5961	17.01200	8.8230	5.1770	23.
1000	6.0890	17.09155	10.04945	3.95055	0.1

! = Estimate per Figure E-1a, where Earth, F = 55, is of Abundance 100

Table E-5b
Distance into the Past That is Observable, Age = 14 Gyrs

For Age = 14 Gyrs, as presented in Table E-5b above, even the most energetic case of observable past distance, that for F = 1,000, has a theoretical limit, about 10 Gyrs, that is less than actual observations have reported [the reported distances based on Hubble - Einstein cosmology].

That Hubble - Einstein cosmology problem is even more severe if the calculations of Table E-5b are performed with no universal decay, as the Hubble - Einstein cosmology contends. The results for that case are presented in Table E-5c below in which the greatest observable past distance is barely 8 Gyrs, quite substantially less than reported observations. That is so even when a greatly more energetic case, F = 3000, is examined. For that extreme the energy is such that gravitation has not yet slowed that matter below, essentially, light speed; which means that its Doppler redshift would be nearly infinite, $z \approx \infty$.

PRIME OBJECTIVE TIME, PRIME OBJECTIVE SPACE

F-Factor Abundance	ChangePoint Time [Gyrs]	2nd Phase Constant-A	Observable Past Distance Back [Gyrs]	Then [Gyrs]	Relative!
1	0.00471	16.59278	3.5544	10.4456	22.
3	0.01413	16.64877	3.7343	10.2657	37.
10	0.04713	16.71513	3.9972	10.0028	63.
32	0.1518	16.79505	4.4214	9.5786	93.
100	0.471	16.89360	5.1720	8.8280	90.
316	1.489	17.01575	6.5490	7.4510	23.
1000	4.710	17.16130	8.08125	5.91875	0.1
! = Estimate per Figure E-1a, where Earth, F = 55, is of Abundance 100					
3000	14.13 [>Age]	n/a	8.15	5.85	

Table E-5c
Table E-5b RE-Calculated with No Universal Decay and Added Extreme Case

Clearly, the tenets of the Hubble - Einstein cosmology fail because they cannot conform to reality as it is already known. Returning to Universal Decay cosmology and the age of the universe being *Age = 30 Gyrs*, that age derives from what is needed to enable observation of redshifts on the order of *z = 10*, as presented in the next section below. It is an estimate because our instrumentation presently limits our ability to observe the past more than the theoretical limit does. Consequently new developments in instrumentation and observation may produce observed redshifts greater than *z = 10*, ones on the order of *z = 12* or more, and therefore require a corresponding increase in the estimated age of the universe.

The present value for the farthest back into the past that it is theoretically possible to observe regardless of the quality of our instrumentation is a little over *27 Gyrs* ago to the time *2 to 3 Gyrs* after the "Big Bang". The travel of the light source and of the observer's home and of the emitted light for that case of the most distant source theoretically observable, all from the time of the "Big Bang" to the present are as shown in Figure E-6, below.

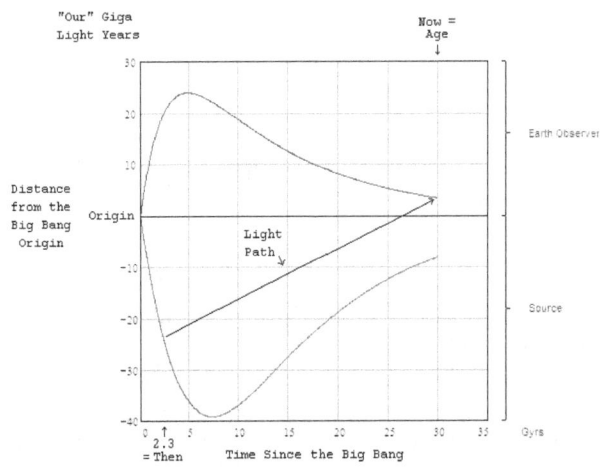

Figure E-6
The Most Distant into The Past Source [-] Observable by an Earth Observer [+] the Two Diametrically Opposite Each Other Relative to the "Big Bang"

E – THE 'BIG BANG' OUTWARD COSMIC EXPANSION

b. Redshifts: Universal Decay and Hubble Doppler

There are two principle causes of the redshifts that we observe: the effect of the universal decay and the Doppler shift due to astral objects' velocity away from us.

The universal decay redshift occurs because we observe ancient light traveling at the speed at which it was originally emitted, a speed significantly larger because less decayed than our present local speed of light. We observe the greater speed as a lengthening of all wavelengths in the light [with no change in frequencies]. The formulation for the universal decay redshift, z_T, of light that was emitted at time $t = T$ after the "Big Bang" and is observed at time $t = now = age$ is as follows.

(E-37) $z_\tau \equiv$ redshift due to the universal decay

$$= \frac{\lambda_{observed} - \lambda_{local}}{\lambda_{local}} = \frac{c(\text{time light emitted}) - c(\text{time now})}{c(\text{time now})}$$

$$z_\tau = \frac{c(t=0) \cdot \varepsilon^{-\left[T/\tau\right]} - c(t=0) \cdot \varepsilon^{-\left[age/\tau\right]}}{c(t=0) \cdot \varepsilon^{-\left[age/\tau\right]}}$$

$$= \frac{\varepsilon^{-\left[T/\tau\right]}}{\varepsilon^{-\left[age/\tau\right]}} - 1$$

The formulation for the Doppler shift due to astral objects' velocity away from us, z_D, is as follows, per standard Hubble - Einstein cosmology.

(E-38) $z_D \equiv$ relativistic redshift due to the Doppler effect

$$= \frac{[1 + v/c]^{1/2}}{[1 - v/c]^{1/2}} - 1$$

The formulation for the universal decay redshift, Equation E-37, is a function of time, not velocity. Equation E-38 can be converted to expressing the Doppler redshift, z_D, in terms of time by using the velocity-as-a-function-of-time expressions for the motion of the astral body products of the "Big Bang" developed earlier above: Equations E-24 and E-27.

For the period from time $t = 0$ through $t = ChangePoint$ the velocity, Equation E-24, is very nearly the then current decaying speed of light. The v/c ratio is very nearly 1.0 so that the redshifts, z_D, are very large, but are also essentially meaningless for any useful purpose. For the period from $t = ChangePoint$ onward the velocity expression $v2$ is Equation E-27, repeated below.

(E-27)
$$v2(t, Age, F) = c(t, Age) \cdot \frac{1}{\varepsilon^{A \cdot (t - ChangePoint)}} \quad \text{meters}/\text{sec}$$

where A and ChangePoint are as given in Table E-5, above.

From Equation E-27 the v/c ratio is:

159

$$(E\text{-}39) \quad v/c = \frac{1}{\varepsilon A \cdot (t - \text{ChangePoint})}$$

and by substituting that into Equation E-38 the expression for the Doppler redshift, z_D, is:

(E-40) $z_D \equiv$ relativistic redshift due to the Doppler effect

$$= \frac{[1 + v/c]^{\frac{1}{2}}}{[1 - v/c]^{\frac{1}{2}}} - 1$$

$$= \frac{\left[1 + \dfrac{1}{\varepsilon A \cdot (t - \text{ChangePoint})}\right]^{\frac{1}{2}}}{\left[1 - \dfrac{1}{\varepsilon A \cdot (t - \text{ChangePoint})}\right]^{\frac{1}{2}}} - 1$$

These two principle causes of redshifts are depicted independently in Figure E-7, below. Of course, the actual observed redshift is the sum of the two. From the figure it is apparent that the Doppler-caused redshifts are quite minor until one is addressing light emitted only at times too early to be observable.

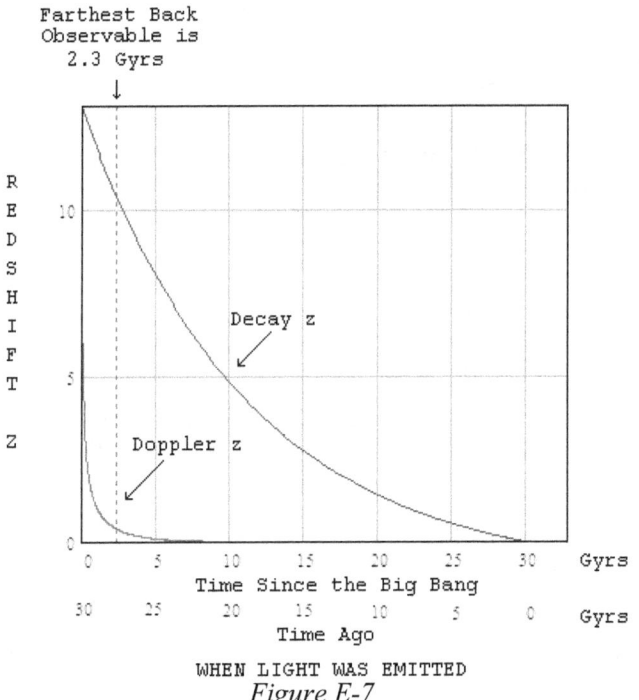

Figure E-7

Redshifts: Caused by Universal Decay and by the Doppler Effect

The figure also makes clear why the age of the universe must be on the order of *30 Gyrs*. That amount of time is needed to include enough time constant periods, τ, that is *11.3373 Gyrs*, each to yield a maximum observable redshift [at *then = 2.3 Gyrs*] of *z > 10* as in the figure. A small number of redshifts at *z = 10* have been reported with some indications of redshifts as high as *z = 12*. Improved instrumentation and techniques may well result in confirmed detection of redshifts at $z > 10$. The value $age = 30\ Gyrs$ is an, at present, conservative best estimate taking into account currently known observational data.

However, the more significant comparison of redshifts is to compare the decay-caused redshift to the redshift associated with the Hubble "Constant". The development is as follows.

(E-41) Where: $c \equiv$ "our" local light speed [the only light speed in Hubble- Einstein cosmology].

$v \equiv$ observed astral body's velocity away from us observers.
$d \equiv$ distance of observed astral body away from us observers.
$z \equiv$ observed redshift.
$H_0 \equiv$ Hubble "Constant".

Then: $H_0 = v/d$ [km/sec/megaparsec]   ["Hubble Law"]

$$v = c \cdot \frac{[z+1]^2 - 1}{[z+1]^2 + 1} \quad [km/sec] \quad \text{[Equation E-40 solved for "v"]}$$

$d = v/H_0$ [megaparsecs]   [solving "Hubble Law" for "d"]

$$= \frac{c}{H_0} \cdot \frac{[z+1]^2 - 1}{[z+1]^2 + 1} \quad \text{[substitute "v" from above]}$$

Three values for H_0 will be illustrated in recognition of the uncertainty of the correct value for it in Hubble - Einstein cosmology:

(E-42) $H1 = 63$ [km/sec/megaparsec] $= 20.5 \cdot 10^6$ [meters/sec/G-Lt-Yr]
        $H2 = 75$ [km/sec/megaparsec] $= 24.4 \cdot 10^6$ [meters/sec/G-Lt-Yr]
        $H3 = 88$ [km/sec/megaparsec] $= 28.7 \cdot 10^6$ [meters/sec/G-Lt-Yr]

and the resulting distances, d, produced by Equation E-41 using c in [meters/sec] and those values of H_0 in [meters/sec/G Lt-Yr] are the same numerical value as the time from the present into the past in Gyrs [for the Hubble - Einstein value of c].

The variation of z vs. t, where $t =$ [time / distance into the past that the observed light was emitted] per the above for the several values of H_0 is depicted in Figure E-8, below, along with the corresponding Universal Decay variation of z vs. t.

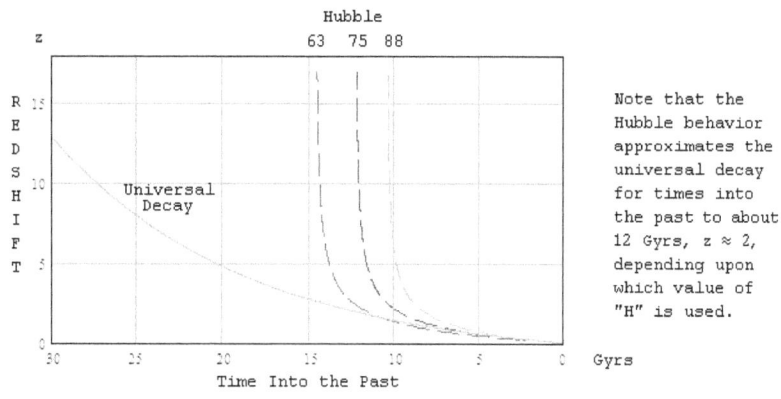

Figure E-8
Redshift Comparison of the Universal Decay and the Hubble - Einstein Concept

The Figure makes clear the reason for Hubble - Einstein cosmology estimates of the age of the universe being in the "mid-teens"; it is the asymptotic behavior of the relativistic Hubble redshift formulation. Unfortunately, the currently more favored value for the Hubble "Constant", $H_0 = 72$ fails to correspond well to the currently favored age of the universe of about *14 Gyrs*. It is also worth noting that the Type Ia Supernovae observations that are the basis for the contended "dark energy" were at redshifts in the range $z = 0.4 - 0.8$, astral objects at up to about *5 Gyrs* into the past.

3 - *The Fate of the Universe*

Although the initial velocities of the "Big Bang" product particles were greater than the there / then escape velocity, as shown earlier above, their velocities now are all well under their escape velocity. We are used to escape velocity being escape velocity -- a simple yes or no proposition. The reason that that is not the case for the overall universe is the effect of the speed of light limitation.

The escape process is the conversion of kinetic energy into gravitational potential energy. If the initial kinetic energy is greater than the maximum possible gravitational potential energy then there will be escape. In the case of a rocket leaving Earth that process is accompanied by the rocket's velocity taking it farther enough away from the Earth that the gravitational effect is reduced in "proper" relation to the process. But the "Big Bang" product particles were not permitted to so travel, their actual velocity being limited to just under light speed as compared to the much larger theoretical non-relativistic velocity at which they would have had to have traveled outward for the accrued distance to correspondingly reduce the gravitational effect in "proper" relation to the process.

The actual velocities and the related escape velocities now at time $t = Age$ for the same distribution of initial "Big Bang" product energies analyzed earlier above is given in Table *E-9*, below. As in the earlier analysis of the initial escape velocity, the present analysis is non-relativistic, using velocities greater than the speed of light rather than letting mass relativistically increase.

For: Universe Age = 30 Gyrs, which means that:
Initial Light Speed = $4.226,895,62 \cdot 10^9$ m/s
Initial Gravitation Constant, $G = 1.870,24 \cdot 10^{-7}$ m³/kg-s²

At Age = 30 Gyrs:

F-Factor	Outward from the Origin of the "Big Bang"*		
	Velocity[m/s]	Distance[G-Lt-Yrs]	Escape Velocity[m/s]
1	$0.00003814 \cdot 10^8$	2.157	$7.418 \cdot 10^8$
3	$0.0001505 \cdot 10^8$	2.400	$7.032 \cdot 10^8$
10	$0.0006230 \cdot 10^8$	2.727	$6.597 \cdot 10^8$
32	$0.002731 \cdot 10^8$	3.205	$6.085 \cdot 10^8$
55 Earth	$0.003700 \cdot 10^8$	3.374	$5.931 \cdot 10^8$
100	$0.01287 \cdot 10^8$	3.980	$5.461 \cdot 10^8$
316	$0.06673 \cdot 10^8$	5.388	$4.693 \cdot 10^8$
1000	$0.3974 \cdot 10^8$	8.034	$3.844 \cdot 10^8$

* = Decayed to Age

Table E-9
Actual Velocities vs. Escape Velocities Now, at $t = Age$

One must immediately conclude that the entire material universe is ultimately destined to collapse back toward the location of its origin, just as a ball tossed straight up from the Earth's surface ultimately returns to its starting point. However, the case of the universe is more complicated than that of the simple ball and there are also two different considerations for the case of the universe: its matter and its radiation.

a. *The Fate of the Universe's Radiation*

The fate of the radiation emitted from sources [primarily astral sources] throughout the universe is very different from the fate of the universe's matter. Most of the universe's radiation continues propagating outward forever, reduced in concentration inversely as the square of the distance from its source, and carrying outward in itself a significant amount of the universe's energy, which energy becomes essentially lost to the remainder of the universe, the universe's matter. That comes about as follows.

(1) Gravitational Redshift and Light Escape

When a particle of mass m climbs in a gravitational field its speed is reduced by the gravitation, which speed reduction reduces its kinetic energy, $\frac{1}{2} \cdot m \cdot v^2$. Conservation is maintained by the kinetic energy loss being replaced by gravitational potential energy increase.

A photon of frequency f has kinetic mass, m_{ph}, [even though it has no rest mass].

$$(E-43) \quad m_{ph} = \text{energy}/c^2$$
$$= h \cdot f / c^2$$

As light, with its kinetic mass, m_{ph}, climbs in a gravitational field, instead of its speed being reduced its frequency is shifted lower [toward the red]. The photon cannot slow down [to correspondingly reduce its kinetic energy as a particle of matter would] because it is constrained by its nature to only travel at light speed, c. Instead the photon frequency, f, decreases, which reduces its energy, $h \cdot f$, its energy of motion that corresponds to kinetic energy.

Then, for a photon to be able to escape from a gravitational field in a manner analogous to escape for a particle of mass, the photon energy, $h \cdot f$, must at least just exceed the depth of the gravitational potential energy pit, $G \cdot M \cdot m_{ph}/R$, that it experiences at the location where the photon is emitted. On that basis the calculation for photon escape would be that the photon frequency must be at least such that

$$(E-44) \quad h \cdot f_{minimum} > G \cdot M \cdot m_{ph}/R$$

however, the photon mass, m_{ph}, depends on f per Equation E-43 so that a directly solvable relationship cannot be obtained on that basis; photon escape is independent of photon frequency.

(2) The "Schwarzschild Radius" and Escape

Astrophysicists treat a quantity called the "Schwarzschild Radius". The line of thought is that the depth of a gravitational potential energy pit from which a particle must climb in order to escape is $G \cdot M \cdot m / R$ where G is the gravitation constant, M is the gravitating mass, m is the mass of the particle attempting escape, and R is the distance from the center of the gravitating mass at which the particle must begin its attempt. To escape, the particle's kinetic energy, $\frac{1}{2} \cdot m \cdot v^2$, must just exceed that potential energy so that, as presented earlier, the escape velocity is $v_{esc} = [2 \cdot G \cdot M / R]^{\frac{1}{2}}$.

From that formulation, as R decreases the required velocity, v increases. Therefore one can calculate a radius, R_S, the "Schwarzschild Radius", for any particular gravitating body mass, M, such that the required escape velocity, v_{esc}, is the speed of light, c, as follows.

(E-45) <u>For Light</u>, a Photon of Mass m_{ph}:

Photon Energy = Gravitational Potential Energy

$$m_{ph} \cdot c^2 = G \cdot M \cdot m_{ph}/R = G \cdot M \cdot m_{ph}/R_S$$

$$R_S = G \cdot M / c^2$$

(E-46) <u>For a Particle</u> of Mass "m":

Kinetic Energy = Gravitational Potential Energy

$$\tfrac{1}{2} \cdot m \cdot v^2 = G \cdot M \cdot m / R$$

$$m \cdot c^2 = G \cdot M \cdot m / R_S \quad \text{[Because KE = TotalE - RestE, then as } v \to c \text{ TotalE} \gg \text{RestE and KE} \to \text{TotalE not } \tfrac{1}{2} \cdot \text{TotalE all because of the relativistic mass increase]}$$

$$R_S = G \cdot M / c^2 \quad \text{[Solve for } R_S\text{]}$$

[The usual presentation, that ignores the effect as in the above note, is $R_S = \underline{2} \cdot G \cdot M / c^2$]

No matter can travel at light speed, therefore matter located at or nearer to the center of the gravitating mass than R_S cannot ever escape. For radiation escape is independent of the frequency and depends only on the distance, R_S.

For the value of R_S for the universe at the instant of the "Big Bang":

· from Equation *E-19* G was $G(0) = 1.870 \cdot 10^{-7}$ m³/kg-s²,

· from Equation *E-16* M was $m_{Universe} = 3 \cdot 10^{49}$ kg, and

· from Equation *E-6* c was $c(0) = 4.226 \cdot 10^9$ meters/sec.

Then, the value of R_S for the universe at the instant of the "Big Bang" was $3.14 \cdot 10^{23}$ *meters* which is *[0.033 G-Lt-Yrs]* and at that instant the actual distance from the center of the "Big Bang" was much less, $d_0 = 4.0 \cdot 10^7$ *meters*. Therefore, at that time, $t = 0$, no matter nor light could escape from the "Big Bang" as already demonstrated and summarized for matter in Table *E-9*, above. The inability of matter to escape did not change thereafter.

However, in its rapid initial expansion at a speed of very nearly $c(0)$, after time approximately $[R_S \div c(0)]$, that is the first two to three million years, the light-source matter of the universe had moved out from the origin to beyond the "Schwarzschild Radius" and whatever radiation was emitted thereafter was free to travel outward forever.

b. *The Fate of the Universe's Matter*

The universe's matter, however, was already embedded in the impossibility of escape and it only remains to investigate its fate.

To this point the material presented has consisted of analytical deductions and reasonable estimates based on fundamentals of physics, the available data, and the tenets of the theories involved. Now, with regard to the fate of the universe's matter, some of what is presented must be limited to "educated" speculation as to the implied future while some still remains reasonable analytical deductions.

Clearly the large range of the present velocities of the universe's matter and of its varied present distances outward from the origin per Table E-9 means that the universe's matter's gradual slowing - direction reversal - inward collapse will result in a wide range of arrival times at the origin of the original expansion of the universe's various portions. [That as juxtaposed to the concept of a universe all together collapsing and then rE-exploding outward in a succession of "big bangs" as has been hypothesized in the not too distant past.] There are, then, several possibilities to be considered.

- Matter arriving at the initial origin crashing into like kind arriving matter.

Recently there have been analyses of what happens when a large asteroid crashes into the Earth, the energies involved and the resulting destruction being immense. One must [speculatively] increase those energies and their results many orders of magnitude to conceive of what would happen at the collision of two planetary bodies, two suns, or two galaxies.

However, the collision would be kinetic and produce great heat, breakdown into particles, and great kinetic energy of those product particles. It would not be as a nuclear fission nor fusion explosion, that is it most likely would not involve a major conversion of matter to energy.

- Matter arriving at the initial origin encountering there nothing but empty space.

Unlike the case of the ball tossed upward from the Earth's surface in which case the Earth is still there when the ball falls back down, it would seem that there is now likely nothing but empty space at the location of the initial origin. A portion of the universe's matter arriving there unopposed would be traveling at high speed [most likely the same [then outward but now inward] speed as was imparted to it in the original "Big Bang" [but as reduced by the Universal Decay of the speed of light]. That body of matter would pass on through and proceed outward again in its own "personal" replay of its earlier role.

Except, that is, that the first time the gravitating matter of the universe was initially all concentrated at the origin whereas the second time that matter is scattered over a large universe volume. The gravitational conditions would be different for the second pass and the escape velocity would also be different. One can only [speculatively] imagine various scenarios for the further travel of that portion of the universe's matter and its peers / partners.

- Matter arriving at the initial origin and there encountering anti-matter.

There are two alternative hypotheses that can be considered with regard to anti-matter creation in the "Big Bang":

· Anti-matter was created, but in a lesser amount than ordinary-matter, and quite shortly thereafter all of the anti-matter mutually annihilated with an equal amount of ordinary-matter leaving essentially no remaining anti-matter and a small remaining amount of ordinary-matter, which is the matter of our universe. In this hypothesis there is no, or negligible anti-matter in today's universe.

This alternative voids the "matter arriving encountering anti-matter" possibility.

· Matter and anti-matter were created in equal, "mirror" amounts and, while most of it promptly mutually annihilated, small equal amounts of each participated in the outward expansion quickly enough to survive. Thus our universe has matter portions [galaxies, galaxy groups, etc.] and similar anti-matter portions and their continued separation in space largely preserves their continued independent existence.

> In this hypothesis matter arriving at the initial origin could encounter anti-matter, which would result in a mutual annihilation. Unlike the kinetic collision case, the result would be an immense amount of energy radiated as gamma rays. Such events involving significant bodies of matter could be the cause of the extremely high energy gamma ray bursts that have been observed but remain un-accounted-for.

· With regard to the two alternative hypotheses:

The first, the present universe essentially lacking anti-matter, would appear to require anti-matter to differ from ordinary-matter other than by being a perfect "mirror". That would appear to conflict with a symmetrical "Big Bang" as required for conservation. Nevertheless, research seeking to discover a discriminating non-symmetrical difference between matter and anti-matter is being conducted without definitive results so far.

The second is what would be expected for a perfectly symmetrical "Big Bang" consistent with conservation. That alternative is pursued in the remainder of this analysis.

The behavior of anti-matter is such that there is no way to discriminate whether a distant astral source is matter or anti-matter: the gravitation is the same; the light emitted is the same.

c. *The Ultimate End of the Universe.*

(1) *The Universal Decay Will Continue*

The universe will continue shrinking to beyond the point of extremely minute, all to no noticeable effect on its internal functioning no matter how small it becomes relative to the size that it is now or originally was.

If one looks back one million years ago, lengths then were greater than the corresponding lengths today by a factor of 1.0009. Clearly the universal decay has little significance in day to day life. In fact, its only significance is for astronomers, because only they can look back into the past far enough to see the effects of the extremely slow decay.

Everything decays proportionately. The ratio at any time, now or in the past or in the future, of the size of things relative to things does not change at all. There is no fixed objective reference by which one could appreciate or notice the decay other than those accessible only to astronomy. Everything is shrinking, but to no noticeable effect. Whatever happens to be left of the universe some inconceivable number of aeons from now will be so extremely minute compared to the size of things in today's universe as to seem to us as nothing. Yet it will operate, function, behave according to the same rules as our universe now, as if it had not decayed at all [again except astronomically], but subject to the events below.

(2) The Universe's Matter Will Gradually Completely Obliterate

Whatever time it takes, eventually all of the universe's matter will be obliterated in mutual annihilations. The process will be a kind of universe "Russian Roulette", annihilations depending randomly on the simultaneous arrival of matter and anti-matter portions of the universe at the location of the initial origin. Such annihilations will extend only to the extent of arriving masses being equal; the un-annihilated surplus of the greater being hurled outward again for another excursion and later chance of annihilation upon its return.

(3) The Universe's Radiation and Energy Will Be Dispersed in Endless Space

All of the radiation and energy of the matter annihilations along with all of the astral and other radiation and energy from the beginning on [including radiation absorbed and subsequently rE-radiated] will disperse outward in space, gradually reddening and so reduced by inverse square dispersion as to eventually amount to essentially nothing.

(4) Nothing to Nothing ...

In the same way as for we humans when our span of life ends it is said, "Ashes to ashes and dust to dust ....", so for the universe it can be said, "It came from nothing and eventually passes on to nothing, to that from which it came".

References

[1] R. Ellman, *The Origin and Its Meaning*, The-Origin Foundation, Inc., http://www.The-Origin.org, 1997. [The book may be downloaded in .pdf files from http://www.The-Origin.org/download.htm].

[2] R. Ellman, *A Comprehensive Resolution of the Pioneer 10 and 11 "Anomalous Acceleration" Problem Presented in the Comprehensive Report "Study of the Anomalous Acceleration of Pioneer 10 and 11" and the Relationship of that Issue to "Dark Matter", "Dark Energy", and the Cosmological Model*, Los Alamos National Laboratory Eprint Archive at http://arxiv.org, physics/9906031. [The value of the error range of a_P, the Pioneer "anomalous acceleration", was adjusted by the researchers in a new recent paper, at http://arxiv.org, gr-qc/0409117 , from ± 0.94 to ± 1.33.]

[3] Freedman, W. L. et al. "Final results from the Hubble space telescope key project to measure the Hubble constant", *Astrophysical Journal*, 533, 47 - 72, (2001).

[4] The fundamental constants are per *The 1986 Adjustment of the Fundamental Physical Constants* as reported by the Committee on Data for Science and Technology [CODATA]. The actual practical precision is limited to one or two significant figures by the nature of the estimates of quantities such as the density and mass of the universe and the universe's expansion velocities and their distribution.

The greater number of significant figures indicated in the data tabulations do not signify greater precision of results. Rather, they are included because they are the actual data used in the calculations before rounding to the real precision, and they are the data used in generating the various graphs. They make the results presented completely reproducible.

[5] Los Alamos National Laboratory Eprint Archive at http://arxiv.org; various papers, e.g. astro-ph/0403025, astro-ph/0403419, astro-ph/0407194, astro-ph/0409485, astro-ph/0410132.

[6] R. Ellman, *How and Why the Universe Began*, Los Alamos National Laboratory Eprint Archive at http://arxiv.org, physics/9904054.

www.ingramcontent.com/pod-product-compliance
Lightning Source LLC
Chambersburg PA
CBHW081428220526
45466CB00008B/2309